NASA/SP—6107–ADD

Reference Mission Version 3.0 Addendum to the Human Exploration of Mars: The Reference Mission of the NASA Mars Exploration Study Team

Bret G. Drake, editor
Lyndon B. Johnson Space Center

June 1998

The NASA STI Program Office ... in Profile

Since its founding, NASA has been dedicated to the advancement of aeronautics and space science. The NASA Scientific and Technical Information (STI) Program Office plays a key part in helping NASA maintain this important role.

The NASA STI Program Office is operated by Langley Research Center, the lead center for NASA's scientific and technical information. The NASA STI Program Office provides access to the NASA STI Database, the largest collection of aeronautical and space science STI in the world. The Program Office is also NASA's institutional mechanism for disseminating the results of its research and development activities. These results are published by NASA in the NASA STI Report Series, which includes the following report types:

- TECHNICAL PUBLICATION. Reports of completed research or a major significant phase of research that present the results of NASA programs and include extensive data or theoretical analysis. Includes compilations of significant scientific and technical data and information deemed to be of continuing reference value. NASA counterpart of peer-reviewed formal professional papers, but having less stringent limitations on manuscript length and extent of graphic presentations.

- TECHNICAL MEMORANDUM. Scientific and technical findings that are preliminary or of specialized interest, e.g., quick release reports, working papers, and bibliographies that contain minimal annotation. Does not contain extensive analysis.

- CONTRACTOR REPORT. Scientific and technical findings by NASA-sponsored contractors and grantees.

- CONFERENCE PUBLICATION. Collected papers from scientific and technical conferences, symposia, seminars, or other meetings sponsored or co-sponsored by NASA.

- SPECIAL PUBLICATION. Scientific, technical, or historical information from NASA programs, projects, and missions, often concerned with subjects having substantial public interest.

- TECHNICAL TRANSLATION. English-language translations of foreign scientific and technical material pertinent to NASA's mission.

Specialized services that complement the STI Program Office's diverse offerings include creating custom thesauri, building customized databases, organizing and publishing research results ... even providing videos.

For more information about the NASA STI Program Office, see the following:

- Access the NASA STI Program Home Page at *http://www.sti.nasa.gov*

- E-mail your question via the Internet to help@sti.nasa.gov

- Fax your question to the NASA STI Help Desk at (301) 621-0134

- Telephone the NASA STI Help Desk at (301) 621-0390

- Write to:
  NASA STI Help Desk
  NASA Center for AeroSpace Information
  7121 Standard Drive
  Hanover, MD 21076-1320

NASA/SP—6107–ADD

Reference Mission Version 3.0 Addendum to the Human Exploration of Mars: The Reference Mission of the NASA Mars Exploration Study Team

Bret G. Drake, editor
Lyndon B. Johnson Space Center
Houston, Texas

National Aeronautics and
Space Administration

Lyndon B. Johnson Space Center
Houston, Texas 77058

June 1998

Available from:

NASA Center for AeroSpace Information
7121 Standard Drive
Hanover, MD 21076-1320
301-621-0390

National Technical Information Service
5285 Port Royal Road
Springfield, VA 22161
703-605-6000

Contents

A1.0 Introduction 1
A2.0 Strategic Modifications 1
   A2.1 Reference Mission 1.0 Launch Strategy 2
      A2.1.1 System Repackaging 5
      A2.1.2 System Mass Reductions 6
      A2.1.3 Modified Launch Strategy 7
   A2.2 Elimination of Initial Habitat Flight 7
      A2.2.1 Volume Augmentation 8
      A2.2.2 Redundancy Considerations 8
   A2.3 Revised Mission Strategy for Version 3.0 9
A3.0 System Design Improvements 10
   A3.1 Incorporation of TransHab Improvements 10
   A3.2 System Improvements 12
   A3.2.1 In-Situ Resource Utilization 12
      A3.2.2 Power Systems 12
      A3.2.3 Science Systems 14
      A3.2.4 EVA Systems 15
   A3.3 Transportation System Improvements 16
      A3.3.1 Earth-to-Orbit Transportation 16
      A3.3.2 Trans-Mars Injection 19
      A3.3.3 Aeroassist 21
      A3.3.4 Descent and Landing 23
      A3.3.5 Ascent 26
      A3.3.6 Trans-Earth Injection 26
      A3.3.7 Launch Packaging 26
A4.0 Summary of Reference Mission Version 3.0 28
A5.0 Revolutionary Next Steps 33
   A5.1 Solar Electric Propulsion 33
      A5.1.1 Electric Propulsion Mission Concept 34
      A5.1.2 Electric Vehicle Concepts 35
   A5.2 Aerocapturing the TransHab 36
   A5.3 Bimodal Nuclear Thermal Rocket (NTR) Propulsion 39
      5.3.1 Bimodal NTR Mission Concept 39
      5.3.2 All Propulsive" Bimodal NTR Option Using TransHab 41
      5.3.3 "LOX-Augmented" NTR Option 42
      5.3.4 Three-Magnum Scenario 43
      5.3.4.1 Mission Content 44
      5.3.4.2 Strategies and Technology Challenges 44
      5.3.4.3 Combination Lander Scenario 45
      5.3.4.4 Split Mission Scenario 45
      5.3.5 All Solar Scenario 48
A6.0 Continuing Work 50

List of Figures

Figure A2-1	Cargo and Piloted Vehicles for Reference Mission Version 1.0	3
Figure A2-2	Reference Mission Version 1.0 Mission Sequence	4
Figure A2-3	Mars Surface Lander and Habitat Aeroshells for Version 1.0	5
Figure A2-4	Habitat Repackaging Strategy	6
Figure A2-5	Historical Space Habitat Pressurized Volume	7
Figure A2-6	Mars Surface Inflatable Habitat Concept	8
Figure A3-1	Payload Capability to 407 km	17
Figure A3-2	Magnum Launch Vehicle	18
Figure A3-3	Schematic of solid core NTR turbopump and power cycle	20
Figure A3-4	NTR stage and aerobraked Mars payload for Version 3.0	22
Figure A3-5	Aeroassist Study Results for Version 3.0	23
Figure A3-6	Entry and Landing Parachute Study Results for Version 3.0	25
Figure A3-7	Launch and Packaging Configurations for Version 3.0	27
Figure A4-1	Reference Mission Sequence for Version 3.0	29
Figure A5-1	Solar Electric Propulsion Mission Concept	34
Figure A5-2	SEP Crew Taxi Concepts	36
Figure A5-3	Conceptual Solar Electric Propulsion Vehicle	37
Figure A5-4	Launch manifest for the Solar Electric Propulsion Vehicle Concept	37
Figure A5-5	Potential Transhab Aerobrake Configurations	38
Figure A5-6	Schematic "Bimodal" NTR System	39
Figure A5-7	"Bimodal" NTR Transfer Vehicle Option	40
Figure A5-8	All Propulsive Bimodal NTR Carrying TransHab	43
Figure A5-9	Three-Magnum Combination Lander Scenario	46
Figure A5-10	Three-Magnum Combination Lander Launch Packages	46
Figure A5-11	Combination Lander Concept on Mars Surface	47
Figure A5-12	Three-Magnum Split Mission Scenario	47
Figure A5-13	Three Magnum Split Mission Launch Packages	48
Figure A5-14	Solar Irradiation at Mars	49

List of Tables

Table A3-1	Mass Reduction Benefits from the Transhab Study	11
Table A3-2	ISRU System Breakdown for Version 3.0	13
Table A3-3	Power System Improvements for Version 3.0	13
Table A3-4	Science Manifest for Version 3.0	15
Table A3-5	Launch Vehicle Requirements	16
Table A3-6	Payload Capability to 407 km	19
Table A4-1	Payload Mass Evolution from Version 1.0 to Version 3.0	28
Table A4-2	Earth Return Vehicle Mass Scrub for Version 3.0	30
Table A4-3	Cargo Lander Mass Scrub for Version 3.0	31
Table A4-4	Piloted Lander Mass Scrub for Version 3.0	32
Table A5-1	Comparison of "Bimodal" NTR to Reference Mission Version 3.0	41

List of Acronyms

ARC	Ames Research Center	LSS	Life Support System
EELV	Evolved Expendable Launch Vehicle	m/s	meters per second
		mt	metric ton
EEV	Earth Entry Vehicle	MAV	Mars Ascent Vehicle
EMU	EVA Mobility Unit	MSFC	Marshall Space Flight Center
EPM	Electric Propulsion Module	MTV	Mars Transfer Vehicle
ERV	Earth Return Vehicle	NTP	Nuclear Thermal Propulsion
EVA	Extra Vehicular Activity	NTR	Nuclear Thermal Rocket
Hab	Habitat	P/C	Physical / Chemical
HEO	High Earth Orbit	PCRV	Pressurized Control Research Vehicle
HLLV	Heavy Lift Launch Vehicle		
IMLEO	Initial Mass in Low Earth Orbit	PMAD	Power Management and Distribution
ISRU	In-Situ Resource Utilization		
JPL	Jet Propulsion Laboratory	psi	pounds per square inch
JSC	Johnson Space Center	PVA	Photovoltaic Array
kg	kilograms	RCS	Reaction Control System
KSC	Kennedy Space Center	SDV	Shuttle Derived Vehicle
kWe	kilowatts electric	SEP	Solar Electric Propulsion
LaRC	Langley Research Center	SP	Special Publication
lbf	pound force	STS	Space Transportation System
L/D	Lift-to-Drag Ratio	TEI	Trans-Earth Injection
LEO	Low Earth Orbit	TMI	Trans-Mars Injection
LeRC	Lewis Research Center	TransHab	Transit Habitat
LFBB	Liquid Fly Back Booster	TPS	Thermal Protection System

Foreword

This Addendum to the Mars Reference Mission was developed as a companion document to NASA Special Publication 6107, "Human Exploration of Mars: The Reference Mission of the NASA Exploration Study Team." The Addendum summarizes changes and updates to the Mars Reference Mission that were developed by the Exploration Office since the final draft of SP 6107 was printed in early 1998.

The Reference Mission is a tool used by the Exploration Team and the exploration community to compare and evaluate approaches to mission and system concepts that could be used for human exploration missions. It is intended to identify and clarify system "drivers", or significant sources of cost, performance, risk, and schedule variation. It does not represent a final or recommended approach to human Mars missions. Several alternative scenarios, including human exploration missions to the Moon, Asteroids, or other targets beyond Earth orbit as well as employing different technical approaches to solving mission and technology challenges, are currently under study by the Exploration Team. Comparing alternative approaches provides the basis for continual improvement to technology investment plans and general understanding of future human exploration missions.

The Addendum represents a "snapshot" of work in progress in support of planning for future human exploration missions through May 1998. Annual publications of revisions to the Reference Mission are planned beginning in late 1998.

Please direct all correspondence and inquiries about this document to:

Exploration Office
Attention: Reference Mission Data Manager
Mail Code EX13
NASA Johnson Space Center
2101 Nasa Road 1
Houston, Texas 77058-3696

A1.0 Introduction

NASA Special Publication 6107 details the work of the Exploration Study Team through the spring of 1994[*]. As described in that report, the primary role of the Reference Mission is two-fold. First, it is used to form a template by which subsequent exploration strategies may be evaluated for consideration as alternate or complementary approaches to the human exploration of Mars. Second, the Reference Mission is intended to stimulate additional thought and development in the exploration community and beyond.

In serving these two purposes, several components of the original Reference Mission (referred to as Version 1.0) have been modified to that which is presented in its current form Mars Reference Mission Version 3.0. The changes are manifested at the strategic, mission, and system levels of development, and augment or improve upon prior work done by NASA's Exploration Study Team. To facilitate and document the ongoing work of the Exploration Team, this addendum will outline the current strategy (as of this addendum's publication date) as well as provide a description of the current systems. Section two of this Addendum provides a brief overview of the changes to the reference approach which are strategic in nature, that is changes which cross many systems and elements. Section three provides a description of improvements to many of the individual systems and elements. Lastly, section four discusses several revolutionary mission approaches and technical options, currently under consideration by the exploration community, which can provide significant improvements in the mission architecture and mass estimates.

A2.0 Strategic Modifications

The original Reference Mission, compiled in the 1993-94 time frame, has been reviewed and improved in many facets of its design. Modifications to that strategy have been made to create a mission offering less risk, lower cost, and better technical approach than previous mission designs. This section will discuss the strategic modifications which have been made to the original Reference Mission, namely alteration of the launch strategy to greatly reduce the required size of the launch vehicle and revision of a mission strategy leading to the elimination of the initial habitat flight.

---

[*] See: www-sn.jsc.nasa.gov/marsref/

A2.1 Reference Mission 1.0 Launch Strategy

Perhaps the biggest assumption of the original Reference Mission centered on the launch system; specifically, a large, yet-to-be-developed launch vehicle was required to place the mission elements into low Earth orbit (LEO). The launch manifest for the mission elements is shown in Figure A2-1. As can be seen, a 200-metric-ton launch vehicle would be required to achieve a human mission in four launches. This scenario consists of three launches for the first trans-Mars injection (TMI) opportunity, followed by three launches at each subsequent opportunity. The first human mission consists of three cargo launches in the first injection opportunity followed by one piloted launch in the following opportunity, each manifested with the specific equipment as shown in the figure.

To graphically illustrate how each of the four launches are conducted to support the first human mission, Figure A2-2 is provided. During the first mission opportunity in 2011, the three cargo vehicles are launched on a nearly Hohmann transfers from Earth to Mars. Reference Mission Version 1.0 was designed such that the Earth Return Vehicle (ERV-1), containing the return habitat, enters a parking orbit about Mars by utilizing an aerocapture maneuver upon arrival at Mars. The other two cargo elements, Cargo-1 and Hab-1, perform an aerocapture followed by aeroentry and landing, delivering the dry ascent vehicle and crew surface habitat to the Martian surface. These components are followed 26 months later (at the next injection opportunity in 2014) by a second surface habitat, Hab-2, piloted by a crew of six. The crew performs an aerocapture followed by aeroentry and landing to the surface in close proximity to the previously deployed surface assets (Cargo-1 and Hab-1). After completion of the 500-day surface mission, the crew ascends to Mars orbit and rendezvous in Mars orbit with the pre-deployed return vehicle (ERV-1).

It was recognized that development of the large 200-metric ton launch vehicle posed a significant technology and development challenge to the mission strategy. Design of the large launcher raises several cost issues (development, new launch facilities, etc.), and the physical size of the launch vehicle is itself a potential limitation to implementing Version 1.0 of the Reference Mission. The requirement of a heavy lift booster was driven primarily by the initial mass to Low Earth Orbit (IMLEO); therefore, an effort was initiated in the fall of 1996 to reduce the required mass and volume of each launch. These efforts were undertaken

CARGO 1
Ascent Vehicle (Dry) & Lander
ISRU Plant & H2 Feedstock
Surface Power System
Pressurized &Rover
3 Tele-operated Science Rovers
Unpressurized Rover
DIPS Cart, Science

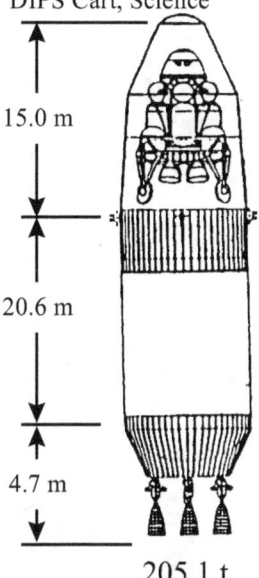

15.0 m
20.6 m
4.7 m
205.1 t

HAB 1
Hab Module and Lander
Surface Nuclear Power
Unpressurized Rover
500 Day Consumables

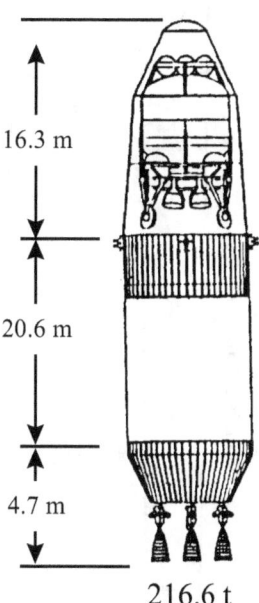

16.3 m
20.6 m
4.7 m
216.6 t

ERV 1
Return Hab LOX/CH4 TEI
180 + 500 Days Consumables

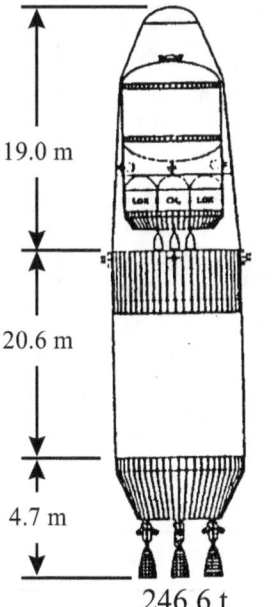

19.0 m
20.6 m
4.7 m
246.6 t

Piloted 1
Surface Hab and Lander
Unpressurized Rover
180 + 500 Days Consumables

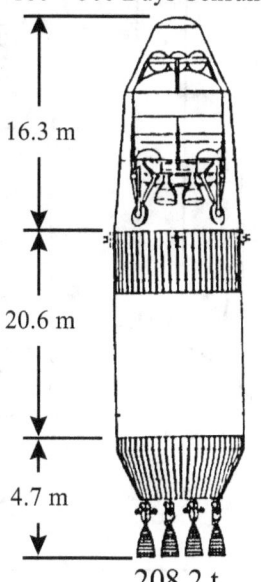

16.3 m
20.6 m
4.7 m
208.2 t

Figure A2-1 Cargo and Piloted Vehicles for Reference Mission Version 1.0.

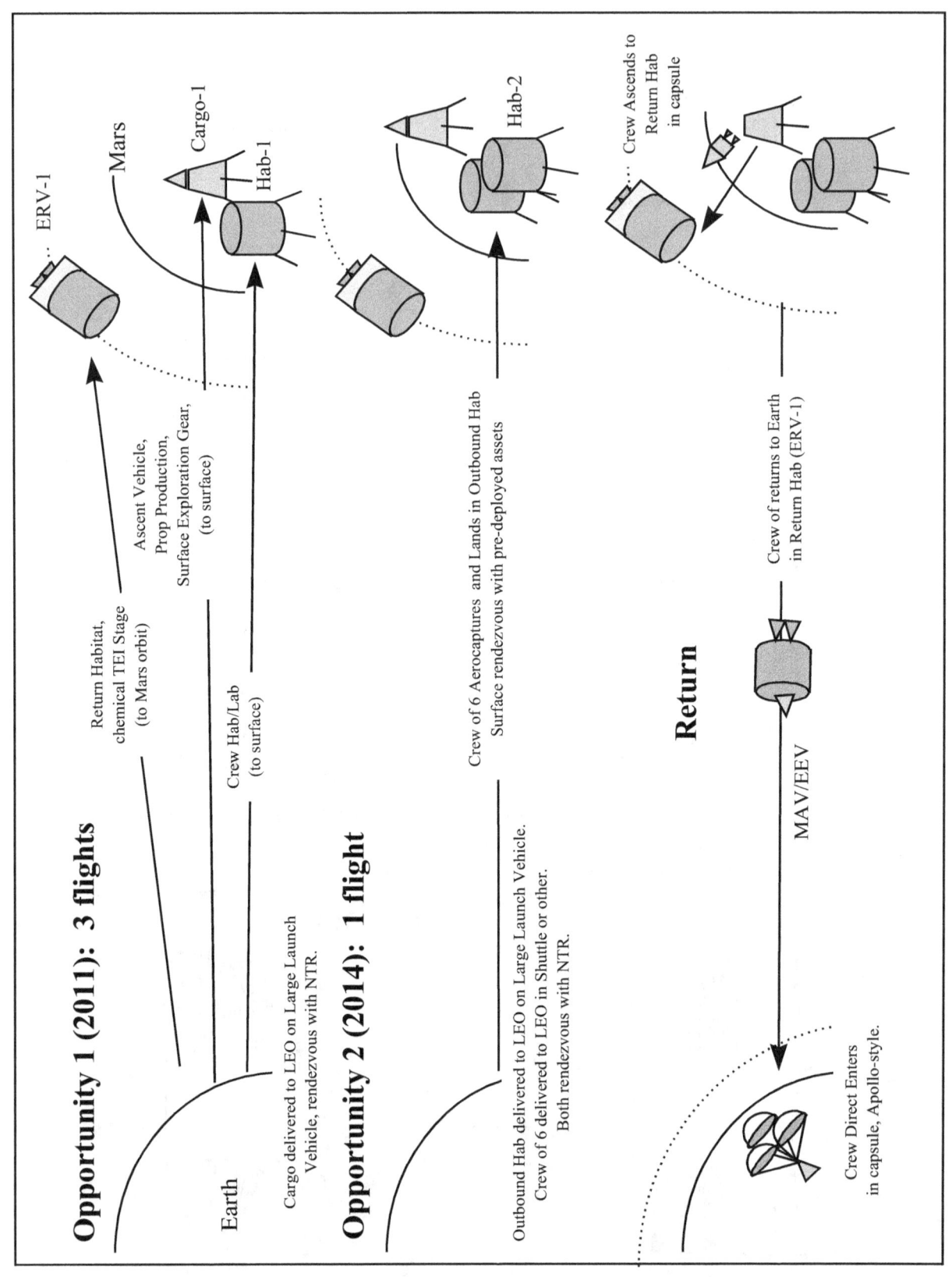

Figure A2-2 Reference Mission Version 1.0 Mission Sequence

while balancing the need to minimize the number of launches to reduce ground launch costs and limit added operational complexity due to Low Earth Orbit (LEO) rendezvous and docking. In order to reduce the size of the launch vehicle, a critical examination of the payloads, in terms of their physical size and mass, was conducted. The goal of this modification was to remanifest the payload elements onto two smaller (80 metric ton class) launch vehicles rather than the single large vehicle.

A2.1.1 System Repackaging

Reducing the physical size of the launch elements is important from many aspects of the launch vehicle design, including reducing the mass of the systems and reducing the aerodynamic loads on the payload shroud. The geometry of the large (10 m diameter) aeroshell for the large launch vehicle, used for both the Mars lander and the surface habitat modules, is given in Figure A2-3. Of particular note is the unused volume between the lander / habitat and the aeroshell.

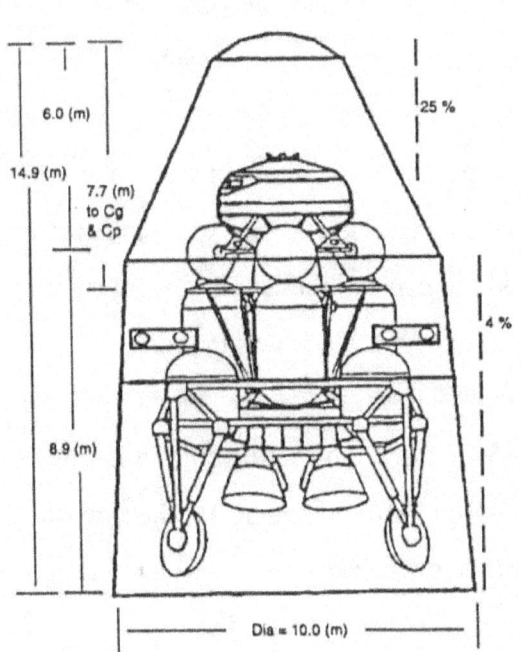

Reference Biconic: 10 (m) Dia by 15 (m) length. l/D = 0.65 At 25° Angle of Attack

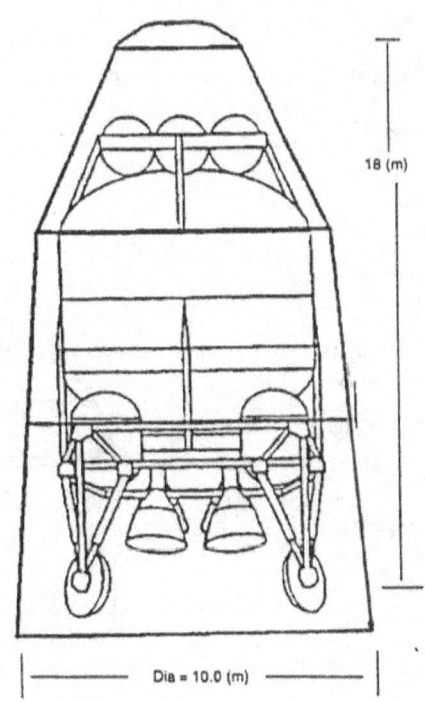

Extended Center Section Biconic 10 (m) Dia by 18 (m) length.

Figure A2-3 Mars Surface Lander and Habitat Aeroshells for Version 1.0.

A proposed solution to this excess volume is shown in Figure A2-4. In this design, the habitat structure is integrated with the Mars entry aeroshell and launch shroud. In addition to reducing the structural mass of the element, the integrated design serves several functions beyond those which were proposed in Version 1.0 of the Reference Mission. Specifically, the integrated habitat / pressure hull with a thermal protection system (TPS):

- serves as both an Earth ascent shroud and Mars entry aeroshell
- eliminates the need for on-orbit assembly / verification of the aeroshell
- allows for stowage in an 80-metric-ton-class launch vehicle.

During the outbound and return interplanetary journeys, Reference Mission Version 1.0 allows for 90 m^3 of pressurized volume per crew member. As can be seen in Figure A2-5, this value is consistent with data from previous space missions. It is desirable to maintain this living quality for the crew despite any subsequent changes which may occur to the original Reference Mission.

A2.1.2 System Mass Reductions

The second step in changing the launch strategy focused on reducing the system masses in order to reduce the mass delivery requirements for the launch system. The payload masses were critically examined, and any duplications were eliminated. In addition, studies were undertaken to scrub the system masses to achieve the required weight savings. The goal of this work was to reduce each payload delivery flight to accommodate the approximate volume and weight limitations of two 80-metric-ton launchers. These mass reductions are discussed in further detail in Section 3.0 of this Addendum.

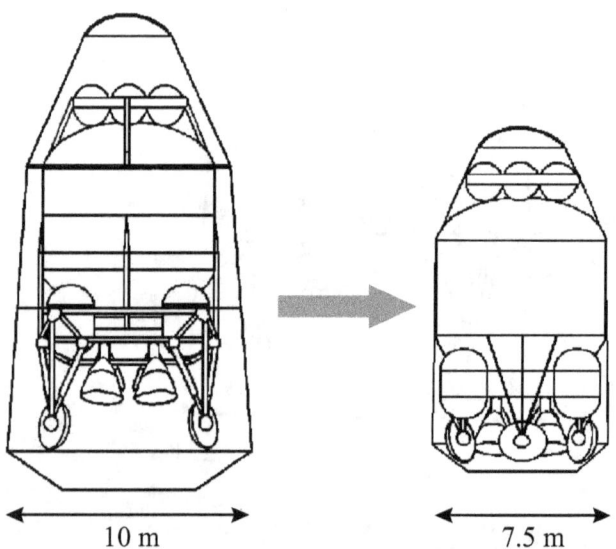

Figure A2-4 Habitat Repackaging Strategy.

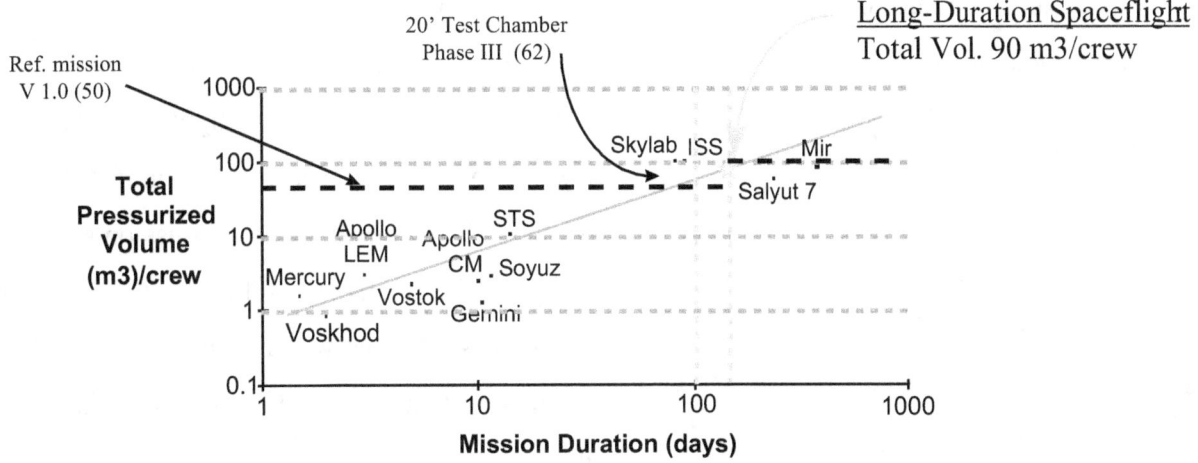

Figure A2-5 Historical Space Habitat Pressurized Volume.

A2.1.3 Modified Launch Strategy

Reduction of the payload delivery flights' mass and volume enables the opportunity to utilize a smaller launch vehicle. This repackaging allows the mission to change from a launch vehicle requiring a 200 metric-ton launch to two individual launches of magnitudes within the envelope of launch systems which can be evolved from current capabilities. This design, delivering the interplanetary propulsion system and cargo into Earth orbit separately, would require one rendezvous and docking operation prior to each outbound journey to Mars. While doubling the number of launches, this strategy eliminates the high costs of developing the large 200 metric-ton launch vehicle of Version 1.0.

A2.2 Elimination of Initial Habitat Flight

While reviewing the original mission strategy, the initial habitat lander (Hab-1) was identified as a launch component which could potentially be eliminated. During the Spring of 1997 a team of engineers at the Johnson Space Center (JSC) investigated a concept of utilizing inflatable structures (known as the TransHab) instead of traditional hard aluminum structures for habitation systems (see Section 3 for more details of this concept). Results of this study demonstrated significant subsystem mass savings for the TransHab concept. Given the significant volume per unit mass increase provided by the inflatable TransHab concept, the attention of the Exploration Team returned to the launch packaging outlined in the original

Reference focusing on techniques of augmenting the surface living volume.

A2.2.1 Volume Augmentation

As noted earlier, a sufficient level of pressurized living volume is critical for crew health maintenance. A TransHab-derived inflatable structure would provide such augmentation, arriving on the Mars surface in the Cargo-1 flight two years before the crew. Pre-plumbed and ready for integration into the life support of the Piloted Crew Lander, the inflatable structure would simply need to be installed by the crew upon arrival, as depicted in Figure A2-6.

The mass of the inflatable module (estimated at 3.1 metric-tons without crew accommodations or life support) could be substituted for the mass of the pressurized rover (5 metric-tons) originally manifested on the Cargo-1 flight. The pressurized rover, deferred to the second Cargo delivery flight, would arrive a few months after the crew and would still be available for the majority of the mission. In essence, the redundancy of the pressurized rover (for the first Mars crew) has been traded for the elimination of an entire Mars-bound habitat flight.

A2.2.2 Redundancy Considerations

The concern that system redundancy would be reduced with the elimination of Hab-1 was mitigated by the redundancy already built into the Reference Mission. For example, several levels of redundancy are present in the mission architecture to address failure of the regenerative lift support system of a habitation module. Four levels of this redundancy are outlined below.

- First level backup - In-Situ Resource Utilization processes generate enough water and oxygen for the entire surface mission to run "open loop."
- Second level backup - The Ascent Vehicle / ISRU plant on Cargo-2 of the subsequent mission, arriving to the surface a few months after the crew, could be used to supply life support rather than for propellant production.

Figure A2-6   Mars Surface Inflatable Habitat Concept.

- Third level backup - The surface could be abandoned for the orbiting Earth Return Vehicle, which has a sufficient food cache to last until the next trans-Earth injection window.
- Fourth level backup - The Earth Return Vehicle (ERV-2) of the subsequent mission, arriving a few months after the crew, would provide an additional refuge for the crew if necessary.

A2.3 Revised Mission Strategy for Version 3.0

The strategic modifications to the Reference Mission described in this section have significantly reduced many of the barriers faced during the formulation of the original approach. The combination of repackaging the mission elements into smaller launch vehicles along with elimination of the initial habitat lander has allowed significant reduction in launch vehicle size, from 200-metric tons down to 80-metric tons, while only introducing two additional flights to the overall launch manifest.

A3.0 System Design Improvements

Point of Contact: Bret Drake/JSC

In order to accomplish the strategic changes discussed in the previous section, improvements to the system designs were required, specifically in terms of system mass reductions. Modifications to the systems were accomplished by the Exploration Study Team (Johnson Space Center, Marshall Space Flight Center, Lewis Research Center, Ames Research Center, Kennedy Space Center, and Langley Research Center), the JSC TransHab study team, and by the Human/Robotic Exploration Team (led jointly by JPL and JSC). Specific improvements include:

- Incorporation of TransHab system designs
- Mass scrub of many of the systems
- Improvements of the transportation system designs

A3.1 Incorporation of TransHab Improvements

Point of Contact: Donna Fender/JSC

In an effort to reduce the cost of human habitation in space, a group from the Engineering Directorate at JSC has been studying an economic and innovative habitation concept based on inflatable structure technology. In the spring of 1997, the improvements associated with the TransHab effort were identified as potential habitat options. Many of the subsystem improvements could be incorporated into the Reference Mission for both the interplanetary and surface phases of the mission.

The Exploration Team has been working to quantify improvements identified in the TransHab study, specifically environmental life support system and structural improvements. It is important to note that the Reference Mission architecture and crew size has remained unchanged with the incorporation of the TransHab option. Some of the masses used by the TransHab team, however, have been scaled to match the duration of the Mars Reference Mission. Advantages of incorporating the TransHab study into the current Mars exploration strategy are manifested primarily in mass reductions. These benefits are provided in Table A3-1. The results are presented for both the Piloted Crew Lander Surface Habitat and the Earth Return Vehicle, and are given in terms of percent changes from the Version 1.0 Reference Mission. Many of the subsystem mass estimates taken from the TransHab studies were of higher fidelity than those previously used by the Mars

Piloted Crew Lander

	"Scrubbed Ref. Mission"*	Version 3.0	Delta		
Habitat Element	33657 kg	19768 kg	-13889 kg	-41%	Mass Reduction due to included systems
P/C LSS	3000	4661	1661	55%	2778 dry plus 1883 fluids
Crew Accommodations	16157	11504	-4653	-29%	Normalized to 680 days
EVA Equipment	1000	969	-31	-3%	Normalized to Six EMUs
Comm/info management	1500	320	-1180	-79%	
Power Dist.	500	275	-225	-45%	
Thermal	2000	500	-1500	-75%	"Mostly" included in LSS Mass
Structure	5500	1039	-4461	-81%	
Crew	500	500	0	0%	
Spares	3500	0	-3500	-100%	Spares accounted for in elements
3kWe PVA/RFC Keep-Alive	1700	1700	0	0%	
Unpressurized Rovers (3)	440	500	60	14%	Payload assumed
EVA Consumables	2300	2300	0	0%	
Crew+EVA Suits**	1300	0	-1300	-100%	Double bookkept - should be deleted
Total Payload Mass	39397 kg	24268 kg	-15129 kg	-38%	
Terminal Propulsion System	4200	4200	0	0%	
Total Landed Mass	43597 kg	28468 kg	-15129 kg	-35%	
Terminal Propellant	10800	7052.2	-3748	-35%	
Aerobrake (15%)	8160	5328	-2832	-35%	
Mars Entry Mass	62557 kg	40848 kg	**-21709 kg**	**-35%**	

*As presented 1/13/97
**Double bookkept - should be deleted

Earth Return Vehicle

	"Scrubbed Ref. Mission"*	Version 3.0	Delta		
Habitat Element	31395 kg	21615 kg	-9781 kg	-31%	Mass Reduction due to included systems
P/C LSS	2000	4661	2661	133%	2778 dry plus 1883 fluids
Crew Accommodations	13021	10861	-2160	-17%	180d Consumables + 500 d add'l food
EVA Equipment	500	485	-15.5	-3%	Normalized to Three EMUs
Comm/info management	1500	320	-1180	-79%	
Power Dist.	500	275	-225	-45%	
30 kWe PVA Power	2974	2974	0	0%	
Thermal	2000	500	-1500	-75%	"Mostly" included in LSS Mass
Structure	5500	1039	-4461	-81%	
Science Equipment	900	500	-400	-44%	Payload assumed
Spares	2500	0	-2500	-100%	Spares accounted for in elements
Jettison Excess Consumables	-6600	-6600	0	0%	
Dock Earth Entry Vehicle & P/L	6900	6900	0	0%	
ERV Mass at TEI	31695 kg	21915 kg	-9781 kg	-31%	
TEI Dry Stage Mass	3500	3500	0	0%	
Earth Return RCS Prop	1100	1100	0	0%	
Earth Return Prop	31800	23231	-8569	-27%	
Aerobrake (15%)	10169	7417	-2752	-27%	
Mars Orbit Insertion Mass	77964 kg	56862 kg	**-21102 kg**	**-27%**	

*As presented 1/13/97

Table A3-1 Mass Reduction Benefits from the TransHab Study.

Exploration Team, which accounts for some of the increases in mass values.

A3.2 System Improvements

A3.2.1 In-Situ Resource Utilization

Points of Contact: Jerry Sanders and Todd Peters/JSC

The fidelity of the In-Situ Resource Utilization system designs were improved during the Fall 1997 design cycle. An improved system design tool was developed which incorporates options and sizing routines for different products (fuels, oxidizers, water for life support, etc.), production processes, cryogenic fluid cooling, and tank sizing. With the increased fidelity of the model, the ISRU system mass estimates were adjusted downward for the plant itself (from 4802k from Reference Mission Version 1.0 to 3941kg) and upward for the hydrogen feedstock (from 4500kg to 5420kg). These estimates reflect a plant that will produce both the ascent propellant and a surface life support system consumables water cache (23 metric tons). The power requirement for the In-Situ Resource Utilization system is driven by both the quantity of products required and the time required to produce the products. Sufficient time for product production is provided such that all required consumables are produced and stored in the surface systems prior to crew departure from Earth. Given these groundrules, the current estimate for the power required is on the order of 45 kWe. Further details describing the mass and power breakdown for the ISRU system are provided in Table A3-2.

A3.2.2 Power Systems

Point of Contact: Bob Cataldo/LeRC

During recent analysis efforts, the surface power system design was revisited in order to obtain mass and cost savings from the original system design. The Reference Mission Version 1.0 surface power system design was based on the reactor technologies developed within the SP-100 program, however with 3-80 kWe closed Brayton cycle (CBC) engines operating at 1100 K. Numerous system trades about this original design were conducted considering power needs, radiation shielding, reactor types, operating temperatures, power conversion technologies, recuperation efficiencies, power distribution voltage, inlet temperature, and number of spare power engines. Updates to the original analysis, including operation at turbine inlet temperatures of 1300 K, enabled a reduction in overall system mass from 14.0 to 10.7 metric tons. Although this assumes a temperature increase of approximately 150 K beyond current Brayton technology, required reactor and fuels technologies remain consistent with those developed within the SP-100 and other DOE/NASA programs. In addition, a first order assessment of the mass impacts of utilizing the same reactor technology as the propulsion system was performed. If feasible and practical, only one development program would then be required for both the propulsion and power systems. A power system based upon a gas-cooled nuclear thermal propulsion engine was estimated to have a mass of

	Subsystem Mass		Subsystem Power	
	Propellants	Life Support	Propellants	Life Support
Compressor	496 kg	193 kg	5645 W	2893 W
Sabatier Reactor	60 kg	50 kg	0 W*	0 W*
Hydrogen Membrane Separator	29 kg	23 kg	288 W	225 W
Methane Water Separator	394 kg	315 kg		1690 W
Pyrolysis Unit	711 kg	1172 kg	3397 W	3911 W
Electrolysis Unit	277 kg		18734 W	
Oxygen Liquefier	43 kg		2215 W	
Methane Liquefier	41 kg		2093 W	
Subtotal	2051 kg	1753 kg	32371	8719 W
Total System	**3,805 kg**		**41,091 W**	

* Reaction is exothermic requiring startup power only (~10 kWe for 1 hour)

Table A3-2 ISRU System Breakdown for Version 3.0

	DRM Version 1.0	DRM Version 3.0	
Reactor type	SP-100	SP-100	Gas cooled (common with propulsion system technology)
Heat transport method	Liquid metal to gas	Liquid metal to gas	Direct
Power conversion	3- 1140 K CBC	3- 1300 K CBC	3- 1300 K CBC
Shield	4-pi, 5 REM/yr @ 360°	4-pi, 5 REM/yr @ 90°, 50 REM/yr @270°	4-pi, 5 REM/yr @ 90°, 50 REM/yr @270°
Distance from Base	2.0 km	2.5 km	2.8 km
Distribution line voltage	2000 V	5000V	5000V
System mass (mt)	12.2	9.3	10.5
Deployment cart (15%)	1.8	1.4	1.6
Total	14.0	10.7	12.1

Table A3-3 Power System Improvements for Version 3.0.

12.1 metric tons. Some of the more salient features of the three designs are shown in Table A3-3. Currently Reference Mission Version 3.0 carries the heavier mass of the gas cooled reactor system.

In addition to the system designs discussed above, other system level trades are being conducted. For instance, additional mass savings could result by using indigenous shielding materials such as soil and/or condensed CO_2. The use of indigenous shielding would minimize the system mass differences shown in the table, since the shield mass is the major component of system mass variation.

These concepts are being evaluated for their impact on the power system design itself as well as other systems that might be required to support this concept, such as, mobile equipment or refrigeration systems. In addition, smaller reactor concepts, such as a 50 kWe power system, have been assessed resulting in a total system mass as low as 5.6 metric tons for an SP-100 based system. These smaller reactor concepts could be used for the initial mission phases, with multiple units providing higher power levels for more robust exploration activities, such as food production.

A3.2.3 Science Systems

Point of Contact: John Gruener/JSC

A review of the science components for the Reference Mission was conducted during the Fall of 1997. The emphasis of this activity was to critically review the science manifest, seeking mass savings. The focus of the review was not to change the science strategy, but merely to seek methods of reducing the science manifest mass estimates. It is desirable to maintain a balance between mass reduction and science content. Due to the time limitations of the study, it was not possible to conduct detailed system designs for the various scientific instruments, instead, emphasis was placed on understanding the current science content as it pertains to previous systems designs and removing any undefined system content (50 kg), unnecessary undefined margins (250 kg), and undefined discretionary science (300 kg). A detailed science manifest of the first human mission for Reference Mission Version 3.0 is shown in Table A3-4.

Surface Science Equipment*	DRM 1.0	DRM 3.0	
Field Geology Package	335 kg	300 kg	35 kg not accounted for
Geoscienc Laboratory Eq.	125 kg	110 kg	15 kg not accounted for
Exobiology Laboratory	50 kg	50 kg	No change
Traverse Geophysical Inst.	400 kg	275 kg	125 kg discretionary margin removed
Geophysical/Meterology Inst.	200 kg	75 kg	125 kg discretionary margin removed
10-Meter Drill	260 kg	260 kg	No change
Meterology Balloons	200 kg	200 kg	Needs better definition
Biomedical/Bioscience Lab	500 kg	500 kg	Needs better definition
Discretionary Science	300 kg	0 kg	Removed
Total	2370 kg	1770 kg	
Cruise Science Equipment*			
Particles & Fields Science	100 kg	100 kg	No change
Astronomy Instruments	200 kg	200 kg	Estimate only
Small Solar Telescope	100 kg	100 kg	No change
Biomedical Instruments	200 kg	200 kg	Needs better definition
Total	600 kg	600 kg	

* NASA Reference Publication 1345

Table A3-4 Science Manifest for Version 3.0.

A3.2.4 EVA Systems

Point of Contact: Robert Yowell/JSC

The EVA consumables estimates for Reference Mission Version 3.0 were improved through the incorporation of a parametric sizing algorithm developed during the TransHab study. In addition, to gain further reductions an assumption was made that consumable mass would only be allocated for two emergency EVAs during transit, allowing for two, eight-hour EVAs, performed by two crew. This resulted in a mass of 48 kg for the transit phases. The transit vehicles also include 195 kg each of EVA support equipment (airlock, airlock systems, EMU spares).

System synergism was also incorporated to gain further mass reductions for the surface phase of the mission. EVA consumable requirements were included in the sizing of the In-Situ Resource Utilization system such that additional oxygen was produced by the ISRU system to provide the necessary consumables for routine surface EVA exploration. Utilizing the locally produced oxygen could save approximately five metric tons for the surface phase of the mission alone (10.9 kg per two-person eight-hour EVA). The current EVA consumable estimates are sufficient for one, eight-hour EVA per week, performed by two crew. Additional consumables for a more robust exploration scenario, including food sticks, batteries, drink bags, visors, etc., but not oxygen which the ISRU provides, have not been included in the EVA estimates for Reference Mission 3.0. Estimates for the additional ancillary consumables, for more robust EVAs, will be incorporated in the next version of the Reference Mission. These changes resulted in a total of 446 kg for the surface phase of the mission. Therefore, the total mass of the EVA consumables is currently estimated at 932 kg versus 3000 kg in Version 1.0. Further examination of the assumptions used to reduce these masses is underway.

The EVA dry mass was slightly reduced from 1000 kg to 940 kg based on inputs from the EVA Project Office at Johnson Space Center. This reflects a mass of 156 kg per suit.

A3.3 Transportation System Improvements

Point of Contact: Steve Richards/MSFC

Re-examination of the performance and design characteristics of the transportation elements for the Reference Mission were led by engineers at the Marshall Space Flight Center with support from the Lewis Research Center, Ames Research Center, Langley Research Center, and Kennedy Space Center. Major modifications to the transportation elements, resulting in Reference Mission Version 3.0, are discussed in this section.

A3.3.1 Earth-to-Orbit Transportation

Points of Contact: Bill Eoff and David Smith/MSFC

Human Mars mission launch costs are driven by initial mass in low-Earth-orbit (IMLEO); launch costs per pound of payload; launch vehicle development costs; and on orbit assembly costs. Earth-to-Orbit (ETO) metrics identified in DRM 3.0 required launch vehicle payload capability of 80 metric tons to minimize on orbit assembly costs and meet payload size requirements. Cost metrics of less than $1000 launch cost per pound of payload and total mission costs of $6B for any launch vehicle development costs and all launch recurring costs have been designated as reasonable starting requirements to drive system designs, see Table A3-5.

Reference Mission 3.0 Payload Requirements	
P/L Diameter:	7.5 m / 24.8 ft
P/L Length:	27.7 m / 91.4 ft
P/L Weight:	80 mt / 176 Klb
Assembly Orbit (28.5 deg)	407 km / 220 nmi
Launch Rate:	6 per year

Table A3-5 Launch Vehicle Requirements

During the design cycle for Reference Mission Version 1.0 numerous configurations were considered and a Shuttle derived vehicle (SDV) with an inline core vehicle was selected. The SDV launch concept barely meets the $6B cost metric for total mission ETO costs because of the high core vehicle costs for Shuttle common hardware. In addition, recent analysis indicated that the SDV configuration exceeded the $1000/lb metric by a factor of two.

Launch vehicle assessments for Reference Mission 3.0 focused on evaluating a core vehicle that is not Shuttle derived to decrease launch costs. Advances in launch vehicle technologies from the Reusable Launch Vehicle (RLV) and Evolved Expendable Launch Vehicle (EELV) programs could make it cost effective to develop a core vehicle that would potentially reduce the $6B ETO cost metric to $2.5B or less per current estimates. This new vehicle concept has been designated as "Magnum" to differentiate from the numerous other past launch vehicle studies. The current Magnum configuration is an inline core vehicle with two attached Shuttle boosters. The payload is aft mounted on the expendable core vehicle; a similar configuration as Titan IV but with over five times the payload capability for one third the launch costs, as shown in Figure A3-1.

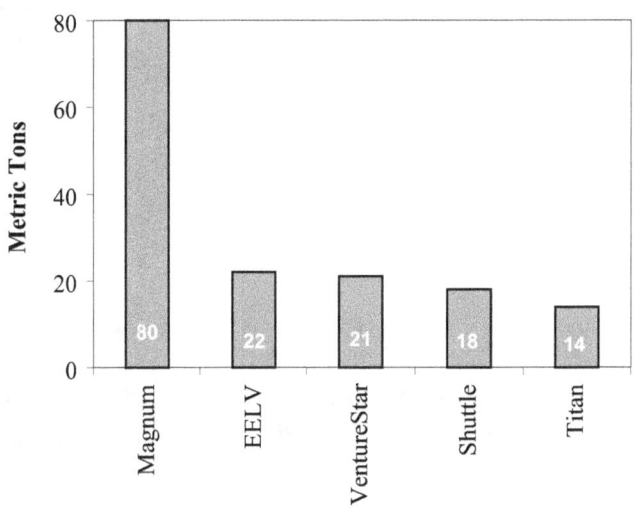

Figure A3-1 Payload Capability to 407 km.

The Magnum vehicle configuration includes a core component which is 8.4 meters (27.5 ft) in diameter, the same as the Shuttle External Tank, to allow common use of Shuttle boosters and launch facilities, see Figure A3-2. By using Shuttle launch facilities and the proposed Liquid Fly Back Boosters (LFBB), recurring costs is estimated to be less than $1000 per pound of payload. A composite shroud is used to protect the payload during ascent and a small kick stage is used for circularizing the orbit. The current design of the Magnum launch vehicle provides a delivery capability of 85 metric tons (188 KLB) to 407 km (220 nmi) orbits at 28.5 degrees inclination or 80 metric tons (176 KLB) to 51.6 degree inclination orbits. See Table A3-6 for additional Magnum performance data.

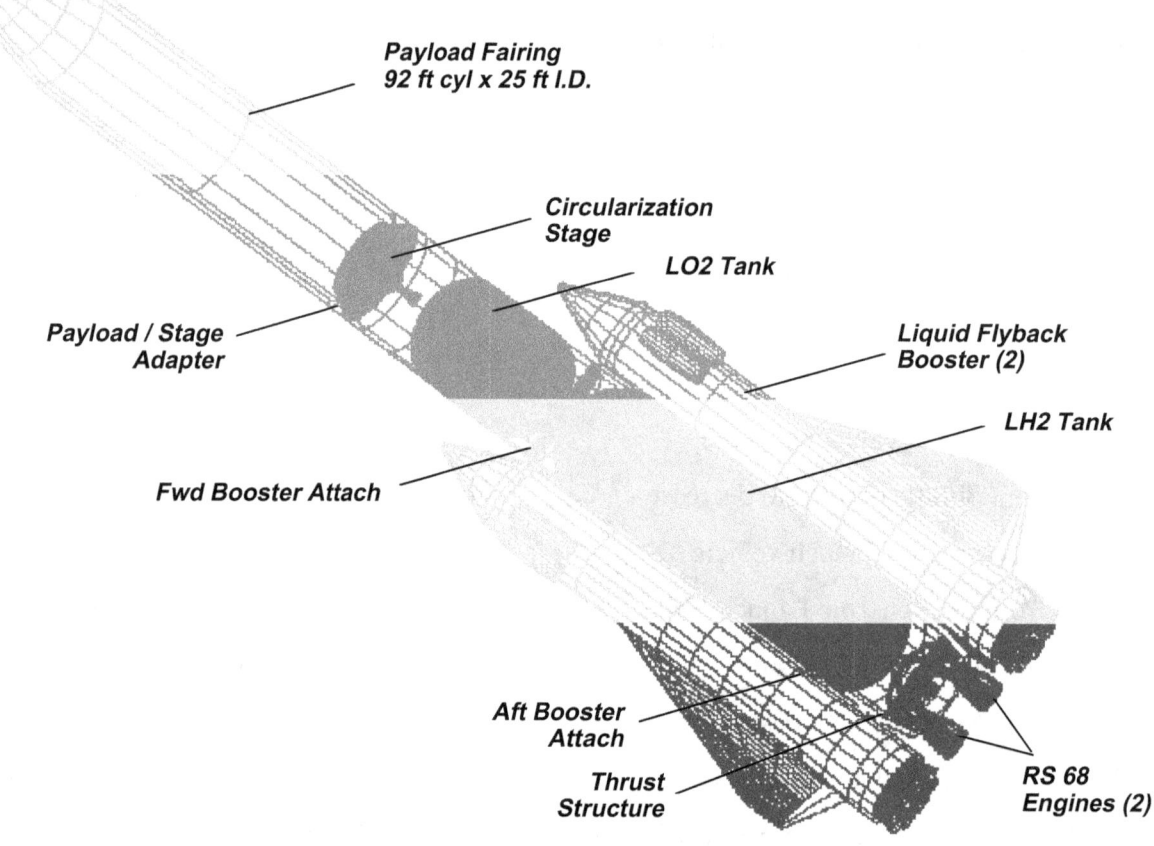

Figure A3-2 Magnum Launch Vehicle.

Technology development and demonstrations for the Magnum launch vehicle concept are driven by the large vehicle size and low life cycle cost requirements. Current evaluations are focused on maximizing the cost-effective application of technologies for engines, valves, composite tanks/structures, and other hardware or facilities under development or projected to be available on other programs such as RLV or EELV. The proposed Magnum technology development program would physically extend these technologies to fit Magnum. Tasks would need to be conducted to demonstrate 8.8 meter (27.5 ft) diameter composite fuel tank manufacturing techniques derived from techniques developed on substantially smaller tanks for RLV. Equivalent

	Performance or Mass (mt)	
	Inclination 28.5°	Inclination 51.6°
Shroud Drop @ 400 Kft	85.4	79.9
Integrated Shroud/Aerobrake		
Payload Only	75.7	70.2
Shroud / Aerobrake Wt.	13.6	13.6
Total Injected	89.3	83.7

Table A3-6 Payload Capability to 407 km.

tasks would be conducted to demonstrate large composite shrouds using the Advanced Grid Stiffened (AGS) composite shroud manufacturing techniques first developed for EELV by the USAF Phillips Lab. Composite structures, propellant ducts and valve technologies would also need to be demonstrated.

Though the Magnum configuration using LFBBs was selected to drive technology developments, the Magnum configuration is still open for assessment of alternate boosters, engines, etc. which would meet requirements.

A3.3.2 Trans-Mars Injection

Point of Contact: Stan Borowski (LeRC)

A high performance trans-Mars injection (TMI) system is required to propel the cargo and piloted spacecraft payloads from their LEO assembly orbits to the desired trans-Mars trajectories and to stay within the mass (~80 metric tons) and payload dimension (~7.6 m diameter x ~28 m length) limits of the Magnum launch vehicle. For Reference Mission 3.0 the solid core nuclear thermal rocket (NTR) was used for the Trans-Mars Injection stage. Other alternatives, such as a Solar Electric Propulsion concept, are currently under investigation as discussed in Section 5.

Conceptually, the NTR engine is relatively simple (Fig. A3-3). High pressure hydrogen propellant flows from the turbo pumps cooling the nozzle, reactor pressure vessel, neutron reflector, control drums, core support structure and internal radiation shield, and in the process picks up heat to drive the turbines. The hydrogen exhaust is routed through coolant channels in the reactor core's fuel elements where it absorbs the energy released by fissioning uranium atoms. The propellant is superheated (to 2,700-3,100 K), and then expanded out a supersonic nozzle for thrust. Controlling the NTR engine during its operational phases (startup, full thrust, and shutdown) is accomplished by matching the turbo pump-supplied hydrogen flow to the reactor power level. Control drums, located in the surrounding reflector region, regulate the number of fission-released neutrons that are reflected back into the core. An internal neutron and gamma radiation shield, containing interior coolant passages, is also placed between the reactor core and sensitive engine components to prevent excessive radiation heating and material damage.

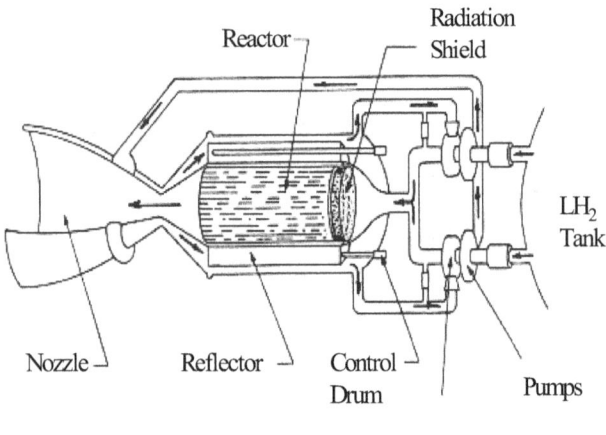

Figure A3-3 Schematic of solid core NTR turbopump and power cycle.

The TMI stage used in Reference Mission 3.0 employs three 15 thousand pounds force (klbf) NTR engines, each weighing 2224 kg, for an engine "thrust-to-weight" ratio of ~3.1. The TMI stage utilizes a "tricarbide" fuel material composed of a solid solution of uranium, zirconium and niobium ceramic carbides. This fuel has been developed and extensively tested in Russia. During reactor tests, hydrogen exhaust temperatures of ~3100 K have been reported for run times of over an hour. For exit temperature in the range of 2900-3075 K, specific impulse values of ~940-960 seconds are estimated for the tricarbide NTR engine assuming a chamber pressure of 2000 psia, a nozzle area ratio of 300 to 1, and a 110% bell length nozzle.

A "common" TMI stage design has been defined for both the Mars cargo and piloted missions. The single tank stage is sized for the energetically demanding 2009 fast transit piloted mission opportunity and is therefore capable of injecting heavier surface and orbital payload elements on minimum energy Mars cargo missions. The NTR TMI stage and its aerobraked Mars payloads are illustrated in Fig. A3-4. The TMI stage LH_2 tank is cylindrical with $\sqrt{2}/2$ ellipsoidal domes. It has an inner diameter of 7.4 meters, an ~20 meter length, and a maximum LH_2 propellant capacity of ~54 tons assuming a 3% ullage factor. The main TMI stage component is the LH_2 tank which is covered by a 2 inch multilayer insulation (MLI) thermal protection system that minimizes propellant boiloff in low Earth orbit to ~0.043 $kg/m^2/day$. Avionics, fuel cell power, storable reaction control system and docking systems are located in the stage forward cylindrical adapter section. Rearward is the stage aft skirt, thrust structure, propellant feed system and NTR engines. The total TMI stage "dry mass" is estimated at ~23.4 metric tons and assumes the use of composite materials for the propellant tank and all primary structures. For the piloted mission, an external disk shield is added to each engine to provide crew radiation protection which increases the stage dry mass by ~3.2 metric tons.

The cargo and piloted Mars spacecraft depart LEO using a "2-perigee burn" Earth departure scenario to reduce gravity losses however single burn departures are also easily accommodated. The total engine burn time for the TMI maneuver is ~35 minutes--about half that demonstrated in the Russian reactor tests. The common TMI stage can inject ~74 and 61 metric tons of payload to Mars on each cargo and piloted mission, respectively. The range of initial mass in Low-Earth Orbit varies from ~135 to 148 metric tons and the overall vehicle length is ~50 meters. Following the TMI maneuver and an appropriate cooldown period, the aerobraked Mars payload and spent TMI stage separate. The storable bipropellant RCS system onboard the TMI stage is then used to perform the final midcourse correction and disposal maneuvers which place the TMI stage onto a trajectory that will not reencounter Earth over the course of a million years.

A3.3.3 Aeroassist

Points of Contact: Jim Arnold and Paul Wercinski/ARC

The purpose of the Summer/Fall 1997 aeroassist study was to develop and end-to-end conceptual design for human aeroassist vehicles consistent with Reference Mission Version 3.0 payloads and configurations. The emphasis of the study was to develop a reliable mass estimate for the aerobrake as well as to provide a better understanding of the technologies required for the eventual development of an aeroassist capability.

The Aeroassist Summer/Fall study used the Design Reference Mission Version 3.0 Piloted Vehicle mission and trajectory for sizing the entry vehicle for aerocapture and descent from orbit. This trajectory had Mars entry speeds of 7.6 km/s, consistent with a 180-day transit in one particular opportunity. A triconic aerobrake shape was chosen as a baseline to accommodate packaging requirements of the payload elements. It was determined that the triconic shape had sufficient lift-to-drag (L/D) capability to meet aerocapture and descent to surface requirements. An L/D = 0.6 was selected for a trim angle of attack of 47 degrees. The aerocapture at Mars was performed without exceeding the 5g maximum deceleration limit which is necessary to maintain crew health and performance during the aerobraking maneuver.

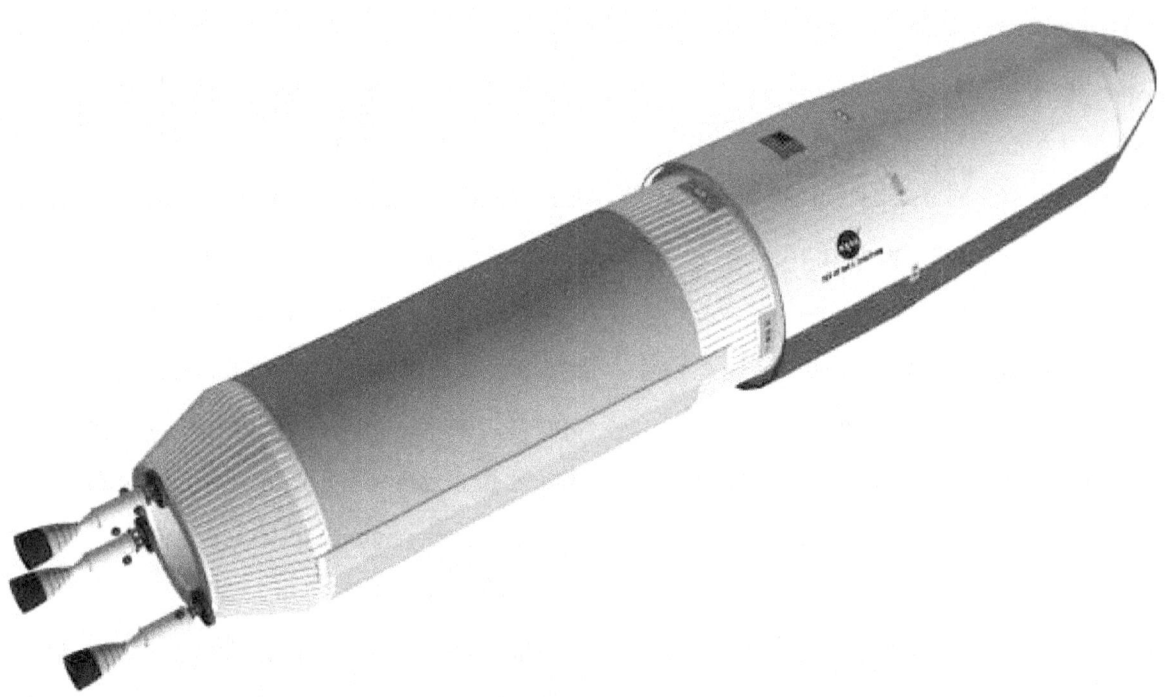

Figure A3-4 NTR stage and aerobraked Mars payload for Version 3.0

Several Navier-Stokes 3-dimensional flowfield solutions were calculated for this shape using appropriate CO_2 chemistry for reacting flows to perform a preliminary thermal protection system (TPS) sizing and trade study for an overshoot trajectory. Turbulent heating estimates were also performed and were identified as a large contributor to uncertainties in predicting heating distribution over the triconic vehicle. Aerodynamic trim was calculated as well and a center-of-gravity location near 49-53% length from the nose was needed for trim. Radiation from the shock layer was also estimated and found to be highly dependent on the reacting gas chemistry models used. Peak heating rates near the nose region were found to between 150-250 W/cm^2. Turbulent flows can result in even higher heating rates downstream. For higher entry velocities, at 8.4 km/s, peak heating rates above 350 W/cm^2 were modeled, but more analysis is needed due to the higher contributions of radiative heating associated with higher entry speeds. Dust erosion effects were also studied and are expected to not be as large of an effect on TPS mass estimates in comparison to turbulent flow or radiative heating issues. Heatshield structure was only estimated by analogy with structure estimates for a Magnum shroud. Heatshield mass estimates (TPS and

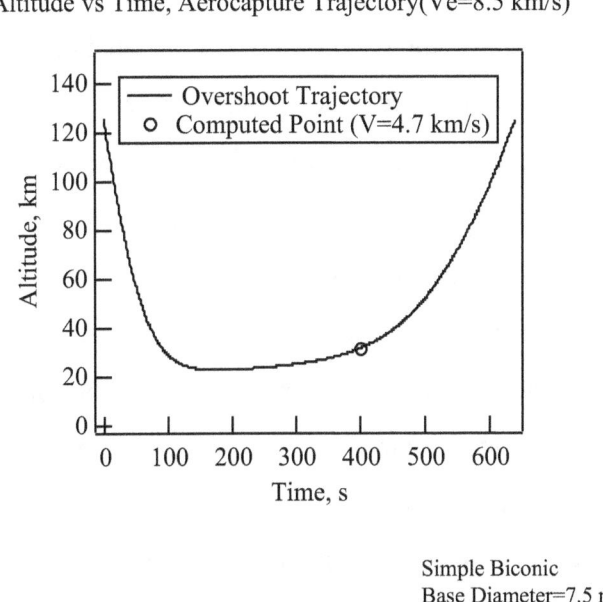
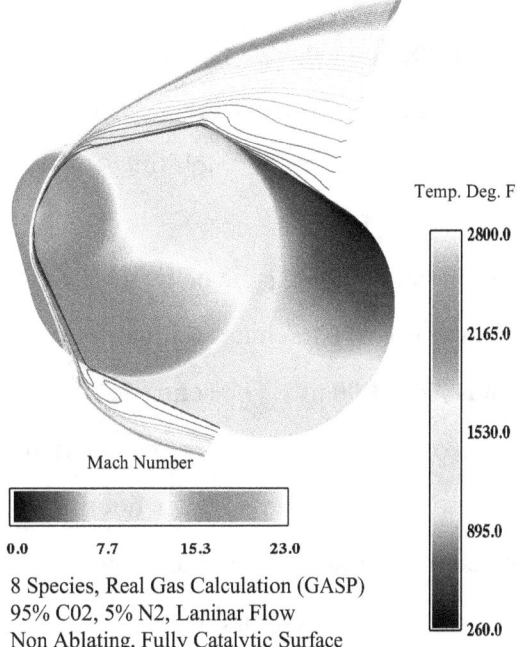

Figure A3-5 Aeroassist Study Results for Version 3.0

structure) yielded mass fractions ranging from 16 – 18% of the total entry vehicle mass. These estimates were used for an entry vehicle carrying 51 metric-tons of cargo. During the aeroassist study, emphasis was not only placed on developing a conceptual approach for human aeroassist, but effort was also devoted to determining key technologies required for aeroassist. The following technology needs the were identified from this study:

- Robust 3D Conceptual Fluid Dynamic code capable of radiating, turbulent, and dusty flows
- Reliable reacting rate/transport and radiation models
- Transition and turbulent models
- Validation methods
- Guidance Navigation &Control. options on approach, L/D > 0.3 guidance capability, terminal descent and landing
- "Human" rated TPS
- 2D TPS sizing tools
- Arc-jets for CO_2 flows
- Flight validation of TPS materials
- High-fidelity integrated design tools supported by local experts across agency.

A3.3.4 Descent and Landing

Points of Contact: Carol Dexter and Larry Kos/MSFC, and Michelle Munk/JSC

Major changes to the descent system for Reference Mission 3.0 include: 1) improved estimates of the descent phase using parachutes, and 2) elimination of the lander mobility requirement.

The descent and landing scheme in Version 1.0 included the use of parachutes with a final landing delta-V of 1000 m/s. The entry to landing phase of the mission was re-examined in Version 3.0 and now includes a higher fidelity method which incorporates mass reductions. Preliminary results were obtained from combining a 3-degree-of-freedom entry simulation and a basic sizing algorithm. In the simulation, the Cargo-1 vehicle, the most massive lander, was deorbited and flown through the atmosphere. Viking-type parachutes were then deployed at about 8 km altitude when the vehicle was traveling roughly 700 m/s. The sizing algorithm was then used parametrically determine the number and size of parachutes and engines required for three different target altitudes. The masses of the parachutes, engines, fuel, and aerobrake were calculated in the sizer and the total vehicle mass was used as the performance metric. The data generated in this analysis are shown in Figure A3-6. A comparison of the new vehicle using the parachute scheme versus the vehicle using the all-propulsive scheme showed a potential savings of ten metric tons. Further analysis of this descent and landing approach includes:

- Verifying the results with an integrated simulation
- Assessment of supersonic deployment of a cluster of large (on the order of 50-m-diameter) parachutes
- Determination of vehicle dynamics
- Consideration of aborts, engine-out situations, and hazard avoidance requirements

Reference Mission Version 1.0 included the capability of the descent system to perform limited surface mobility. This capability was provided so that the two surface habitats could be brought together and essentially "docked" to integrate the livable volume for the crew. With the deletion of the initial habitat, the descent system surface mobility mechanisms are not required, thus significantly reducing the complexity of the descent system design.

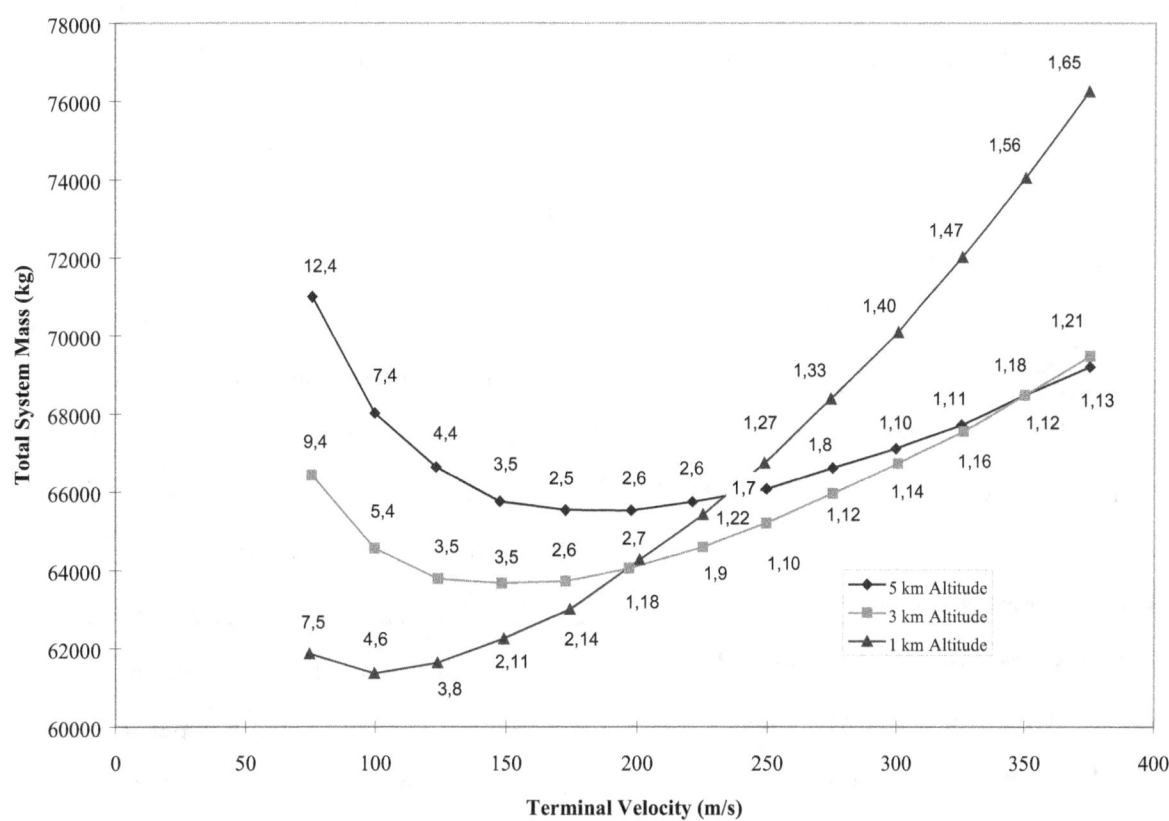

Figure A3-6 Entry and Landing Parachute Study Results for Version 3.0

Given the improvements in the entry and landing scenario and deletion of the surface mobility requirements, the descent system was refined. The descent system employs four RL10-class engines modified to burn LOX/CH$_4$. These are used to perform the post-aerocapture circularization burn and the final 632 meters per second of descent velocity change after parachute deployment. The descent engines are also used for orbital correction maneuvers during the transit from Earth to Mars, the orbit adjust and trim maneuvers after aerocapture, and the de-orbit burn prior to the atmospheric entry and landing. Architecture definitions for the engine include:

- Specific impulse of 379 seconds
- Mixture ratio of 3.5
- Chamber pressure of approximately 600 psi
- A nozzle area ratio of approximately 400
- Thrust level of approximately 15,000 lbf.
- Additional requirements are that the engines be capable of throttling and gimbaling although specific ranges for these parameters have not been determined.

The descent system for Reference Mission Version 3.0 is capable of placing approximately

40 metric tons of cargo on the surface. The dry mass of this system is approximately 4.9 metric tons requiring 11 metric tons of propellant.

A3.3.5 Ascent

Points of Contact: Carol Dexter and Larry Kos/MSFC

The major modification of the ascent stage for Reference Mission Version 3.0 is the incorporation of a common descent/ascent propulsion system approach. The ascent stage propulsion system shares common engines and propellant feed systems with the descent stage. This eliminates the need for a separate ascent propulsion system reducing the overall mass and subsequent cost. These common engines are the same RL10-class engines modified to burn LOX/ CH_4 as the descent stage. These engines perform with an average specific impulse of 379 seconds throughout the ascent maneuver. The ascent propulsion system will require approximately 39 metric tons of propellant to accomplish the approximately 5,625 meters per second of velocity change required for a single-stage ascent to orbit and rendezvous with the previously deployed ERV. The structure and tanks needed for this propellant and the other attached hardware elements have a mass of 4.1 metric tons, including the mass of the engines but not the crew capsule.

A3.3.6 Trans-Earth Injection

Point of Contact: Larry Kos/MSFC

Improvements were also made to the Trans-Earth Injection (TEI) stage for Reference Mission Version 3.0. The TEI stage uses two RL10-class engines modified to burn LOX/ CH_4, similar to the descent stage. These engines perform with an average specific impulse of 379 seconds throughout the TEI maneuver. The TEI stage requires approximately 29 metric tons of propellant and has a dry mass of 5.9 metric tons.

A3.3.7 Launch Packaging

Point of Contact: Larry Kos/MSFC

During 1997 the Exploration Transportation Team led my the Marshall Space Flight Center performed a packaging and launch configuration analysis of the Reference Mission Version 3.0 payload elements. The focus of the packaging analysis was to determine the overall launch sequence and payload dimensions to ensure that the mission elements would fit within the overall payload dimensions and launch strategy of the Magnum Launch Vehicle. An overview of the launch packaging analysis is provided in

Figure A3-7. As can be seen in the figure, the overall launch sequence of the mission elements begins approximately 97 days prior to the opening of the Trans-Mars Injection window. This timeline is driven primarily by the launch processing of the payload elements and launch vehicle. For this analysis, 30 days were allotted for element processing between launches. A more thorough analysis of the ground processing is currently underway to determine a better estimate for the processing timeline.

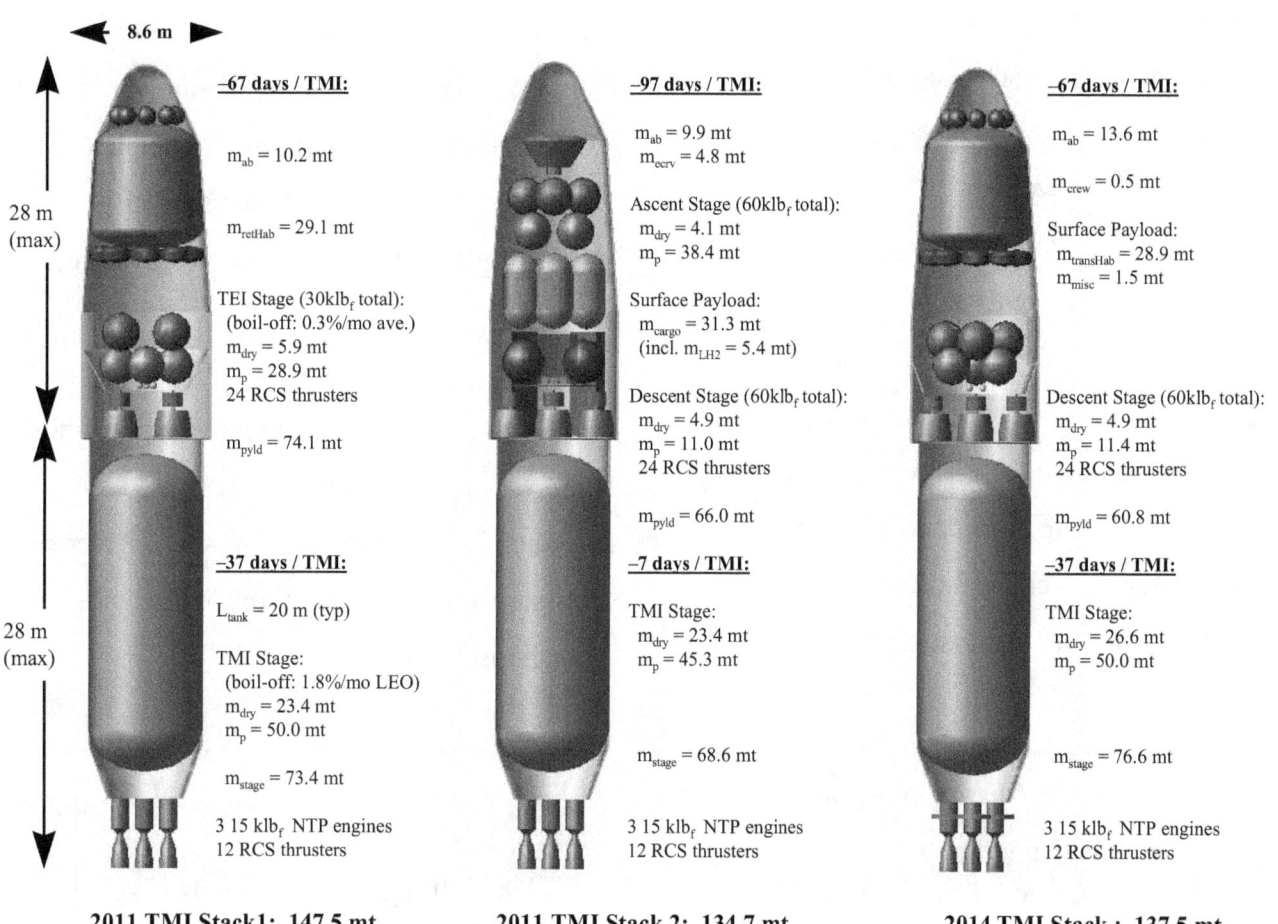

Figure A3-7 Launch and Packaging Configurations for Version 3.0

A4.0 Summary of Reference Mission Version 3.0

From the work of the original Reference Mission (Version 1.0), the strategy for the human exploration of Mars has evolved from its original form to one of reduced system mass, use of a smaller, more reasonable launch vehicle, and use of more current technology. The steps which have been taken by the Exploration Team are motivated by the need to reduce the mass of the payload delivery flights, as well as the overall mission cost, without introducing additional mission risk. By eliminating the need for a large heavy-lift launch vehicle and deleting the redundant habitat delivery flight in Version 3.0, two launches from the Earth were eliminated. The net result is a current Version 3.0 Reference Mission which requires an injected mass of approximately one-half that of the 1993/94 Reference Mission (Table A4-1).

The modifications which have been made to the Reference Mission have resulted in significant reductions in total initial mass in low-Earth-orbit without significantly altering the overall mission architecture. A complete overview of the current Reference Mission Version 3.0 architecture is provided in Figure A4-1.

A comparison of the mass breakdown for the various flights are provided in Table A4-2 through Table A4-4. The masses of Reference Mission 1.0 and 3.0 are provided for comparison.

Reference Mission Version	1.0	3.0
First Opportunity:		
Cargo Lander (Cargo-1)	90,190 kg	66,043 kg
Habitat (Hab-1)	90,598 kg	N/A
Earth Return Vehicle-1 (ERV-1)	131,374 kg	74,072 kg
Second Opportunity:		
Crew+Habitat 2 (Piloted-1)	89,980 kg	60,806 kg
TOTAL	402 mt	201 mt

Table A4-1 Payload Mass Evolution from Version 1.0 through Version 3.0

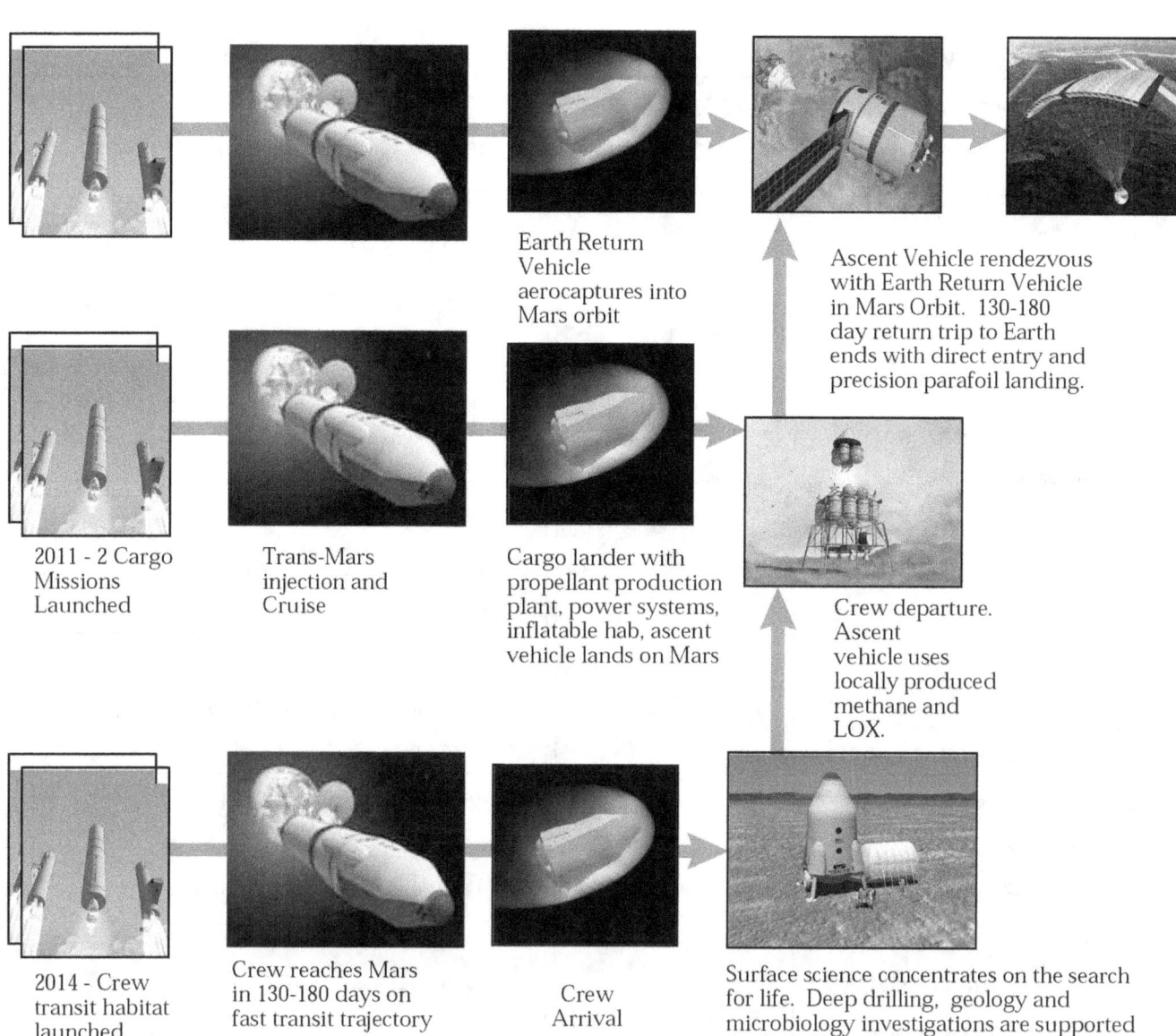

Figure A4-1 Reference Mission Sequence for Version 3.0

Earth Return Vehicle	Reference Version 1.0		Final Version 3.0		
Habitat Element	51974	kg	26581	kg	
Life Support System	6000	kg	4661	kg	TransHab Study 7/97
Crew Accomm. + Consumables	22500	kg	12058	kg	TransHab Sizer from K. Kennedy
Health Care	2500	kg	0	kg	Included in Crew Acc. & Cons.
EVA equipment	4000	kg	243	kg	TransHab Sizer from K. Kennedy
Comm/info management	1500	kg	320	kg	TransHab Study 7/97
30 kw PVA power system	3474	kg	3249	kg	B. Cataldo 12/97 & TransHab Study 7/97
Thermal Control system	2000	kg	550	kg	TransHab Aerobrake Study 11/97
Structure	10000	kg	5500	kg	Scaled from Boeing 7.6 x 16.2 Mars Hab
Science equipment	900	kg	600	kg	12\97 Scrub
Spares	3500	kg	1924	kg	Rodriggs/Munk 12/97
SUBTOTAL	56374	kg	29105	kg	
Excess consumables		kg	-7392	kg	Jettison 560 days of food before TEI
Crew & Samples from Surface	500	kg		kg	Crew and Payload from surface
ERV MASS AT TEI	63274	kg	27042	kg	
TEI stage drymass	5200	kg	4806	kg	MSFC update 1/98
Propellant mass	52000	kg	28866	kg	MSFC update 1/98
Earth return RCS propellant	0	kg	1115	kg	MSFC update 1/98
Aerobrake	17300	kg	10180	kg	ARC update 1\98
TOTAL TEI MASS	120474	kg	61829	kg	
TOTAL MOI MASS	131374	kg	74072	kg	
NTR Propulsion System	28900	kg	23400	kg	
Shadow Shield	0	kg	0	kg	
TMI Propellant	86000	kg	50000	kg	
TOTAL INITIAL MASS	246274	kg	147472	kg	

Table A4-2 Earth Return Vehicle Mass Scrub for Version 3.0

Cargo Lander 1	Reference Version 1.0		Final Version 3.0		
Earth Entry/Mars Ascent Capsule	5500	kg	4829	kg	JSC 12/97 Updates (X-38, ACRV, EVA)
Ascent stage dry mass	2550	kg	4069	kg	MSFC update 1/98
ISRU plant	4802	kg	3941	kg	T. Peters update 1/98; CH4, O2, H2O
Hydrogen feedstock	4500	kg	5420	kg	T. Peters 1/98
PVA keep-alive power system	300	kg	825	kg	B. Cataldo 11/97
160 kw nuclear power plant	12498	kg	11425	kg	B. Cataldo 12/97
1.0 km power cables, PMAD	1900	kg	837	kg	B. Cataldo 12/97
Communication system	820	kg	320	kg	TransHab 7/97
Pressurized Rover	15500	kg	0	kg	Delayed
Inflatable Laboratory Module		kg	3100	kg	Derived from TransHab Study 7/97
15 kwe DIPS cart	1100	kg	1500	kg	DRM v 2.0
Unpressurized rover	440	kg	550	kg	PSS ESDB May 1991, p.486
3 teleoperable science rovers	1320	kg	1500	kg	DRM v 2.0
Water storage tank	1220	kg	150	kg	T. Peters 1/98
Science equipment	3800	kg	1770	kg	J. Gruener Update 12/97
TOTAL CARGO MASS	**56250**	**kg**	**40236**	**kg**	
Vehicle Structure	0	kg	3186	kg	MSFC Update 1/98
Terminal propulsion system	4670	kg	1018	kg	MSFC Update 1/98
TOTAL LANDED MASS	**60920**	**kg**	**44440**	**kg**	
Propellant	11970	kg	10985	kg	MSFC Update 1/98
Forward Aeroshell	17300	kg	9918	kg	ARC Update 1/98
Parachutes and mechanisms		kg	700	kg	4 parachues (to go with 4 engines)
TOTAL ENTRY MASS	**90190**	**kg**	**66043**	**kg**	
NTR Propulsion System	28900	kg	23400	kg	
Shadow Shield	0	kg	0	kg	
TMI Propellant	86000	kg	45300	kg	
TOTAL INITIAL MASS	**205090**	**kg**	**134743**	**kg**	

Table A4-3 Cargo Lander Mass Scrub for Version 3.0

Crew Lander	Reference Version 1.0		Final Version 3.0		
Habitat element 2	53400	kg	28505	kg	
Life Support System	6000	kg	4661	kg	TransHab Study 7/97
Health Care	2500	kg	0	kg	Included in Crew Accommodations
Crew Accommodations	22500	kg	12058	kg	TransHab Sizer from K. Kennedy
EVA equipment	4000	kg	243	kg	TransHab Sizer from K. Kennedy
Comm/info management	1500	kg	320	kg	TransHab Study 7/97
Power	500	kg	3249	kg	B. Cataldo 12/97 & TransHab Study 7/97
Thermal	2000	kg	550	kg	TransHab Aerobrake Study 11/97
Structure	10000	kg	5500	kg	Scaled from Boeing 7.6 x 16.2 Mars Hab
Science	900	kg		kg	12/97 scrub
Spares	3500	kg	1924	kg	Rodriggs/Munk 12/97
Crew	500	kg	500	kg	6 - 183 lb people
3 kw PVA keep-alive power	1700	kg	0	kg	Included above
Unpressurized rover 3	440	kg	550	kg	DRM v 2.0
EVA consumables		kg	446	kg	TransHab Sizer from K. Kennedy
EVA suits		kg	940	kg	R. Yowell est. 12/97 - 156 kg/suit
TOTAL PAYLOAD MASS	56040	kg	30941	kg	
Vehicle structure		kg	3186	kg	MSFC Update 1/98
Terminal propulsion system	4670	kg	1018	kg	MSFC Update 1/98
TOTAL LANDED MASS	60710	kg	35145	kg	
Propellant	11970	kg	11381	kg	MSFC Update 1/98
Forward Aeroshell	17300	kg	13580	kg	ARC Update 1/98
Parachutes and mechanisms		kg	700	kg	4 parachues (to go with 4 engines)
TOTAL ENTRY MASS	89980	kg	60806	kg	
NTR Propulsion System	28900	kg	23400	kg	
Shadow Shield	3300	kg	3200	kg	
TMI Propellant	86000	kg	50000	kg	
TOTAL INITIAL MASS	208180	kg	137406	kg	

Table A4-4 Piloted Lander Mass Scrub for Version 3.0

A5.0 Revolutionary Next Steps

The Mars Reference Mission described in NASA Special Publication 6107, as modified by the updates described in the this addendum, provides a general framework for the human exploration of Mars. Since the original framing of the Reference Mission, other approaches have been brought forward as potential mission and technology options. These approaches, currently being analyzed by the Exploration Team, seem to be promising alternatives for accomplishing the primary objectives set forth in the original mission plan. The major mission alternatives currently under investigation include:

- A Solar Electric Propulsion (SEP) option for performing the Earth departure phase of the mission
- An approach for capturing the inflated TransHab into Mars orbit
- Derivatives of the Nuclear Thermal Rocket concept which produces both propulsive thrust and continuous power
- Techniques for minimizing launch mass perhaps to meet a three-Magnum launch scenario, and
- All solar power scenarios.

A5.1 Solar Electric Propulsion

Points of Contact: Kurt Hack and Leon Geffert/LeRC, and Jeff George/JSC

Many different approaches have been developed utilizing both solar electric and nuclear electric propulsion as a method of transporting both cargo and crew to and from Mars. These approaches focused on how an electric vehicle could be utilized to perform all of the major trajectory phases of the mission, including trans-Mars injection, Mars orbit capture, and trans-Earth injection. Although highly efficient from a propellant utilization standpoint, the relatively high power levels required to achieve fast-piloted trips generated two major challenges: 1) The vehicles were very large requiring significant on-orbit assembly and/or deployment, and 2) The technology requirements were significant (lightweight, multi-megawatt-class nuclear or solar powerplants; efficient and durable thrusters scaled to power levels on the order of 500 kWe). These two significant challenges eliminated the electric propulsion vehicle as the primary propulsion concept for human Mars missions.

During the Spring of 1997 an alternate concept of utilizing electric propulsion was proposed by the Lewis Research Center. After examining the payload delivery requirements of the Reference Mission, it was determined that a compromise approach would be to utilize electric propulsion to perform the bulk of the trans-Mars injection, rather than all mission phases. This would minimize the disadvantages of previous approaches while still providing significant mission benefits.

A5.1.1 Electric Propulsion Mission Concept

The solar electric approach currently under investigation by the Exploration Team utilizes the high efficiency of electric propulsion where it provides the most benefit - boosting cargo out of the Earth's gravity well. The overall mission strategy for the electric propulsion option is fundamentally the same as that of the Reference Mission: two cargo elements are launched in the first mission opportunity, followed by a piloted vehicle in the subsequent opportunity. The only major difference occurs in the replacement of the nuclear thermal TMI stage with a solar electric "tug" and small chemical kick stage. An overview of the mission concept is shown in Figure A5-1.

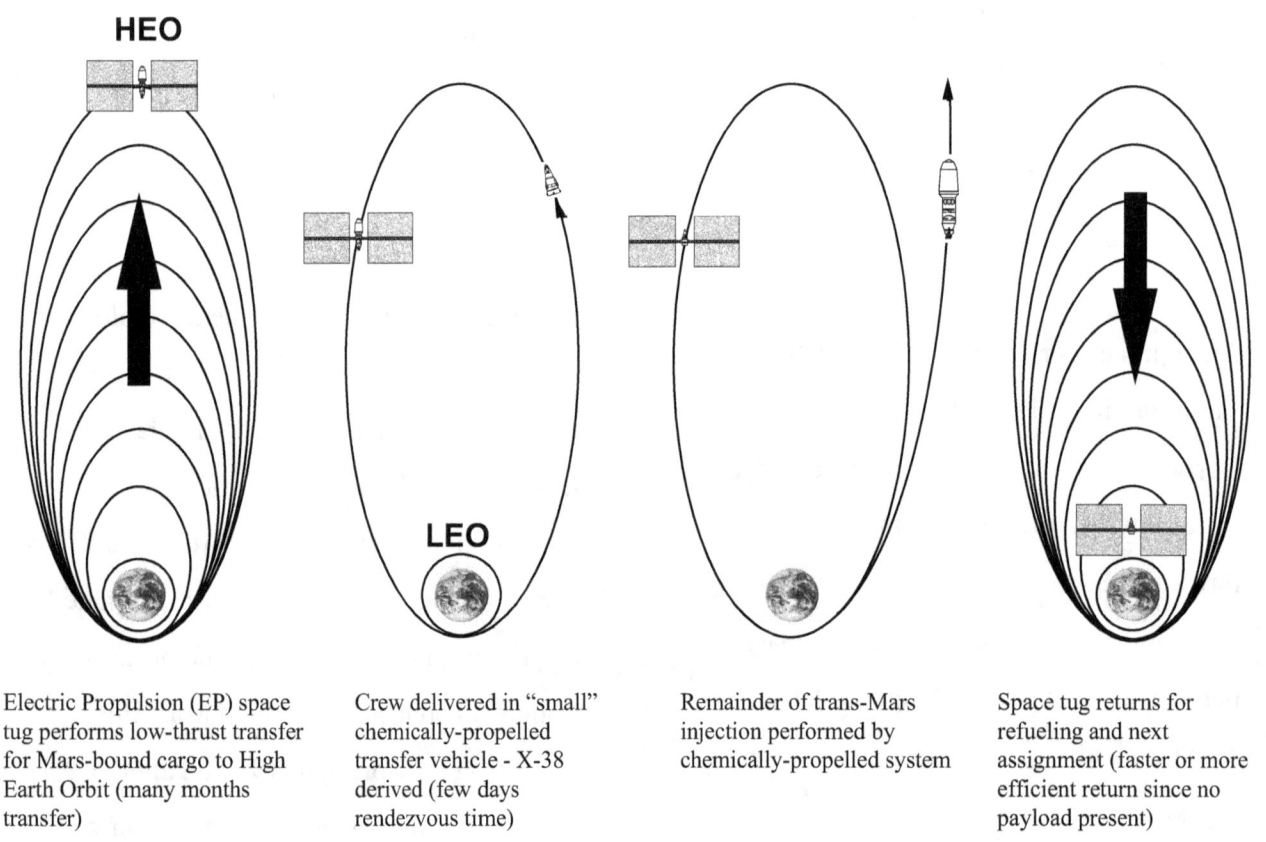

Electric Propulsion (EP) space tug performs low-thrust transfer for Mars-bound cargo to High Earth Orbit (many months transfer)

Crew delivered in "small" chemically-propelled transfer vehicle - X-38 derived (few days rendezvous time)

Remainder of trans-Mars injection performed by chemically-propelled system

Space tug returns for refueling and next assignment (faster or more efficient return since no payload present)

Figure A5-1 Solar Electric Propulsion Mission Concept

Injection of cargo and piloted mission elements to Mars begins with the electric propulsion spiral phase. Due to the inherent high specific impulse at low thrust characteristics of electric propulsion, mission elements cannot be directly injected toward Mars via a traditional short impulsive burn. Orbital energy is instead continuously added over a period of approximately nine months, with the vehicle and payload following a spiral trajectory from an initial circular low Earth orbit (LEO) to a final elliptical high Earth orbit (HEO). A small chemical stage is then used to provide the final injection of the mission cargo toward Mars. The now-unloaded solar electric vehicle then returns to LEO to await a repeat sortie of the piloted vehicle element in the succeeding mission opportunity.

Delivery of the crew to Mars requires a slight modification to the front-end of the mission. As with the cargo missions, the electric propulsion vehicle is used to boost the piloted vehicle, sans crew, into a high Earth orbit. The crew is not transported in the vehicle during this phase for two primary reasons. First, during the spiral boost phase of the mission, the vehicle traverses the harsh Van Allen radiation belts many times - far too excessive for piloted missions. Second, the spiral phase takes several months to perform, significantly increasing the exposure of the crew to the debilitating effects of zero-gravity. Rather than employing countermeasures, these effects are minimized by delivering the crew in a high speed taxi to the piloted vehicle after it has been boosted to the final high Earth departure orbit. After a short rendezvous and checkout period, the piloted vehicle, like the previous cargo vehicles, is injected to Mars with a small chemical stage.

A5.1.2 Electric Vehicle Concepts

Vehicle concepts for the electric propulsion option are currently under investigation by the Exploration Team. During the selection and analysis process, emphasis is being placed on developing a concept which can be deployed easily, do not require significant advancements in technology, and is a low cost approach. Conceptual vehicle designs for the crew taxi and solar electric vehicle are shown in Figures A5-2 and A5-3. The concepts shown are still under investigation and will continue to evolve as advancements in the analysis are made. A summary of the mission mass estimates for the solar electric vehicle concept are provided in Figure A5-4.

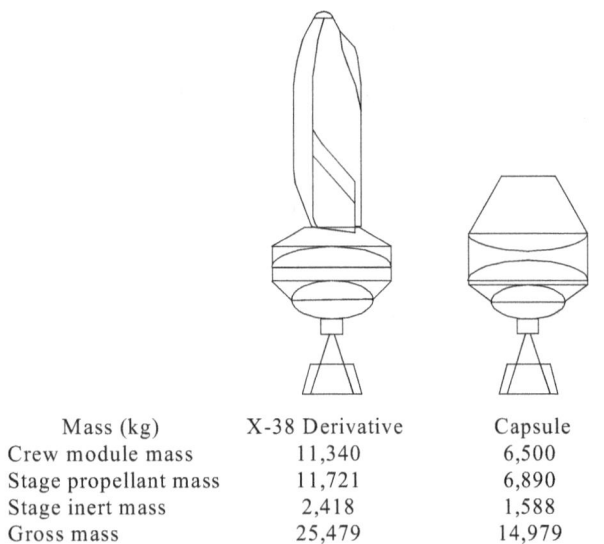

Mass (kg)	X-38 Derivative	Capsule
Crew module mass	11,340	6,500
Stage propellant mass	11,721	6,890
Stage inert mass	2,418	1,588
Gross mass	25,479	14,979

Figure A5-2 SEP Crew Taxi Concepts.

A5.2 Aerocapturing the TransHab

Point of Contact: Bill Schneider/JSC

The other major mission option currently under investigation is the approach of aerocapturing the inflated TransHab into Mars orbit. During the Fall of 1997 a "Skunk Works" study team composed of experts from the Johnson Space Center, Langley Research Center, Ames Research Center, and Marshall Space Flight Center, conducted a study of the TransHab aerocapture concept. The goals of the study were to design a lightweight aeroshell system capable of capturing the inflated TransHab into Mars orbit and to determine the best system for crew return to Earth. Aeroentry and landing were not considered during this study and were left for follow-on analysis.

Two aeroshell concepts were analyzed during the study: The Ellipsled, which uses the structure from the Magnum launch vehicle shroud (requiring no on-orbit assembly); and the Spherical Dome, which is Shuttle-launched and assembled (see Figure A5-5). In order to estimate the total system mass, the analysis included investigations of the entry flight dynamics, thermal protection system, structural design, and assembly operations.

Analysis conducted showed that both the Ellipsled and the Spherical Dome could accomplish an aerocapture at Mars with positive margins. However, a number of factors has led to a selection of the slender shape concept. Results from the study indicated that the mass fraction (ratio of the mass of the aerobrake to the mass of the aerobrake and payload) ranged from 14.6% to 15.5%. These results closely matched those from the previous aeroshell analyses (see section A3.4), indicating reliability in the mass estimates. Aerobrake mass fractions of this magnitude provide significant mission advantages by reducing the total mass required for the mission.

Analyses of the TransHab Aerobrake concepts are still in progress. Several factors

Figure A5-3 Conceptual Solar Electric Propulsion Vehicle

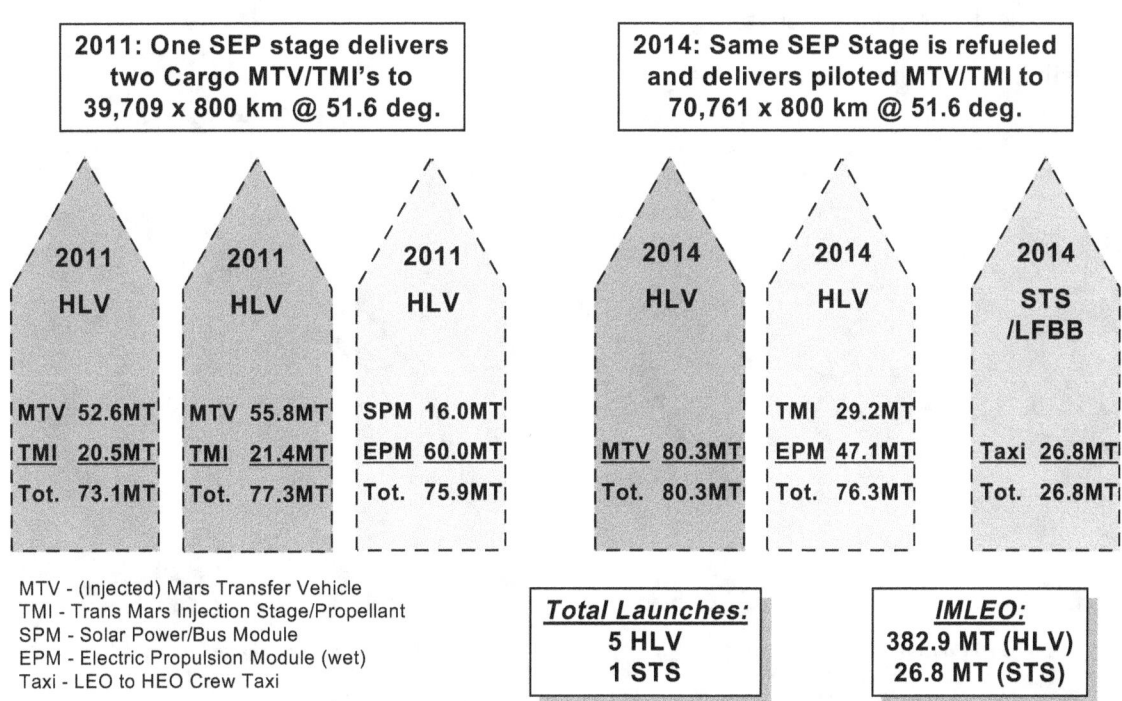

Figure A5-4 Launch manifest for the Solar Electric Propulsion Vehicle Concept.

remain to be investigated to complete the study, including:

- Modifications to the TransHab. The initial effort focused on aerocapturing the habitat originally designed by the JSC team in the Spring of '97. Modifications to the TransHab, including structural modifications to operate on the surface of Mars, and the addition of a crew flight deck, were not addressed.

- Entry and Landing. The initial study focused only on the aerocapture phase of the mission, and did not address the issues associated with the entry and landing phases, such as static and low-speed dynamic stability, parachute deployment, terminal engine requirements, or landing accuracy. If it is not feasible to land the inflatable TransHab in its current configuration, modifications and additional vehicle elements may have to be introduced into the architecture.

- Assessment of the impacts of decreasing the Earth-Mars transit times from 200-days to 180-days to be consistent with previous analysis performed by the Ames Research Center.

TransHab-to-Aeroshell Attachment. Detailed structural design of the attachments between the TransHab and the aeroshell will serve to further refine the mass estimates of the system, and must address packaging, deployment, and accessibility issues.

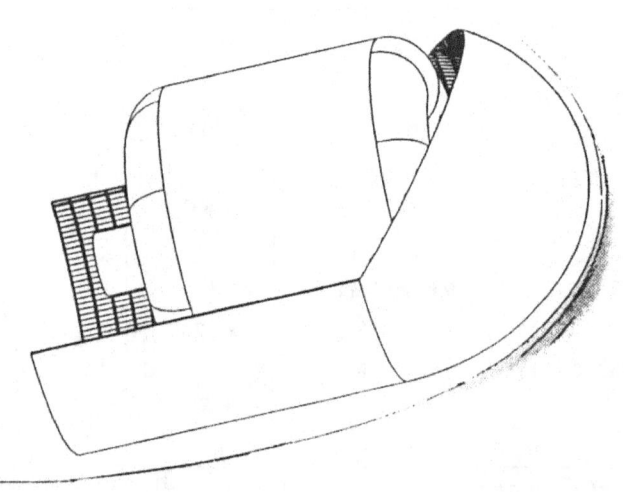
Ellipsled Concept

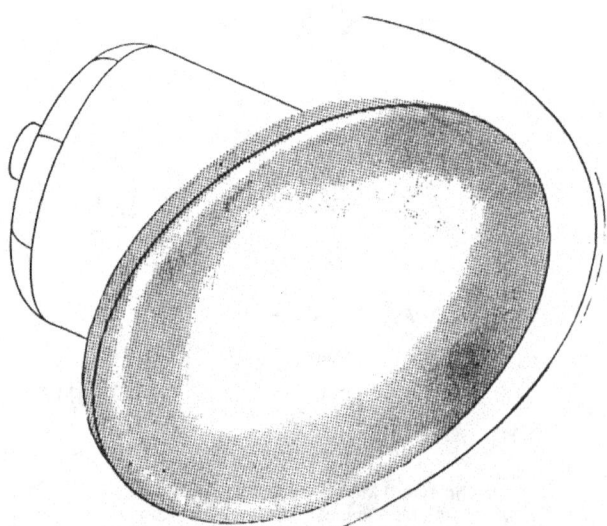
Sphere Dome Concept

Figure A5-5 Potential TransHab Aerobrake Configurations.

A5.3 Bimodal Nuclear Thermal Rocket (NTR) Propulsion

Point of Contact: Stan Borowski/LeRC

Although most of the current work is focused on the Solar Electric concept, the NTR approach is being maintained for comparison. The solid core NTR propulsion system represents a "rich source of energy" in that it contains substantially more uranium-235 fuel in its reactor core than it consumes during its primary propulsion maneuvers. By configuring the NTR engine as a "bimodal" system, abundant electrical power can also be generated for a variety of spacecraft needs. During power generation, the reactor core operates in essentially an "idle mode" with a thermal power output of ~100 kilowatts. The reactor thermal energy is subsequently removed and routed to a turbo-alternator-compressor Brayton power conversion unit using a helium-xenon working fluid, as shown in Figure A5-6. A space radiator system rejects waste heat and also reduces decay heat propellant loss following propulsive burns.

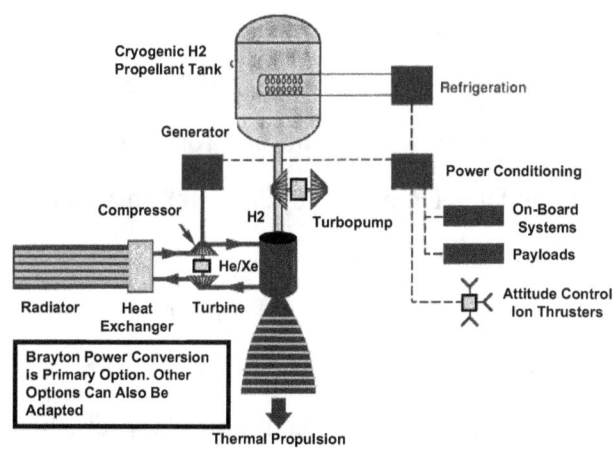

Figure A5-6 Schematic "Bimodal" NTR System

5.3.1 Bimodal NTR Mission Concept

An option to Reference Mission Version 3.0 that utilizes bimodal NTR transfer vehicles in place of the expendable NTR stages is being evaluated. A common "core" stage, used on cargo and piloted vehicles alike, is outfitted with three 15 klbf bimodal NTR engines capable of providing up to 50 kilowatts of electrical power (kWe) using any two engines The bimodal core stage is not jettisoned after the TMI maneuver but remains with the cargo and piloted payload elements providing midcourse correction (MCC) propulsion and all necessary power during transit. Near Mars, the bimodal stage separates from the aerobraked payloads and performs its final disposal maneuvers. A key difference between Reference Mission 3.0 and the bimodal option is the absence of the aerobraked LOX/methane (CH_4) TEI stage which is replaced by an "all propulsive" bimodal NTR-powered Earth Return Vehicle (ERV) illustrated in Figure A5-7.

The bimodal stage LH_2 tank is slightly shorter than the expendable TMI stage tank at 19 meters and has a maximum LH_2 propellant capacity of ~51 tons with a 3% ullage factor. A turbo-Brayton refrigeration system is located in the forward cylindrical adaptor section to eliminate LH_2 boiloff during the lengthy (~4.3 year) ERV mission. A 12 kWe Brayton refrigeration system is included to remove the ~100 watts of heat flux penetrating the 2 inch MLI system in low-Earth-orbit where the highest heat flux occurs. Enclosed within the conical aft radiator section of the bimodal core stage is a closed Brayton cycle (CBC) power conversion system employing three 25 kWe Brayton rotating units (one for each bimodal reactor) which operate at ~2/3 of rated capacity, thus providing an "engine out" capability. The turbine inlet temperature of the He-Xe working gas is ~1300 K and the total system specific mass is estimated to be ~30 kg/kWe.

A mass comparison of the bimodal NTR transfer

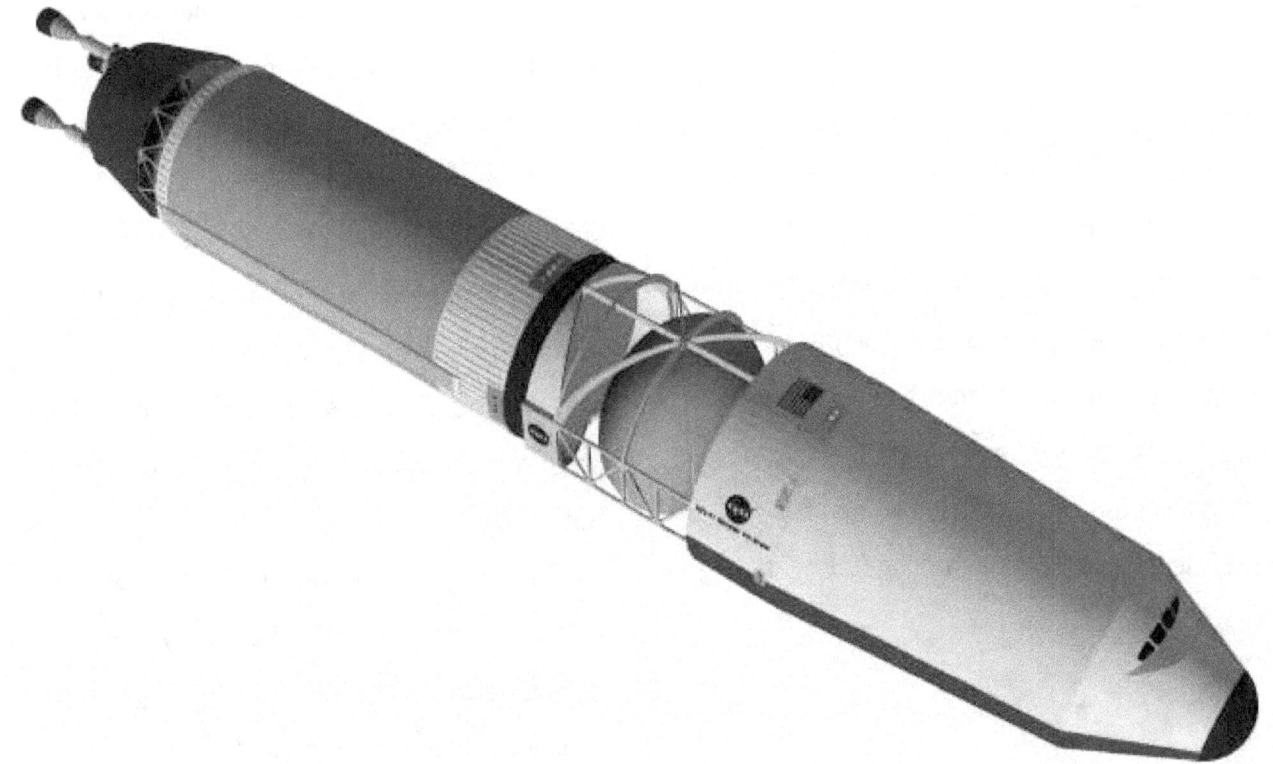

Figure A5-7 "Bimodal" NTR Transfer Vehicle Option

Magnum Launch	Flight Element	2011 Cargo Lander		2011 ERV *		2014 Crew Lander		Totals	
	Mission Type	DRM	Bimodal	DRM	Bimodal	DRM	Bimodal	DRM	Bimodal
#1	Payload - Surface/"In-Space" - Transportation	66.0 - 40.2 - 25.8	65.0 - 40.2 - 24.8	74.1 - 29.1 - 45.0	25.5 - 25.5 	60.8 - 30.9 - 29.9	56.4 - 28.4 - 28.0	200.9 - 100.2 - 100.7	146.9 - 94.1 - 52.8
	"In - Line" Propellant/Tankage (LH_2 &/or LOX)	-	-	-	20.1	-	5.3	-	25.4
#2	NTR TMI stage ("Modified" DRM uses "bimodal" NTRs)	68.6	73.6	73.4	79.0	76.6	79.0	218.7	231.6
	Total :	134.6	138.6	147.5	124.6	137.5	140.7	419.6	403.9
	# Magnums	2	2	2	2	2	2	6	6

* 2011 ERV mission using "bimodal" NTRs for MOC and TEI is lighter than DRM by ~23 t and eliminates DDT&E and recurring costs for LOX/CH4 TEI stage, also recurring cost for 30 kWe PVA and aerobrake.

** Common "Bimodal" NTR TMI stage provides 50 kWe power capability to the ERV, Crew and Cargo lander missions. Also supplies MCC burns for these missions. For cargo lander, the "Bimodal" stage refrigeration/heat rejection systems can be used to cryocool 4.5 t of "seed" LH_2 and dump "waste heat" from 15 kWe DIPS power cart.

Table A5-1 Comparison of "Bimodal" NTR to Reference Mission Version 3.0

vehicles and the Reference Mission Version 3.0 vehicles is shown in Table A5-1.

The mass values assume a "2-perigee burn" Earth departure scenario. Overall, the bimodal approach has a lower "three-mission" initial mass than Reference Mission 3.0. In addition, the bimodal approach can reduce the operational complexity of the mission (eliminates solar array deployment/retraction) as well as eliminating the need for an aerobrake and injection stage for the Earth Return Vehicle.

5.3.2 All Propulsive" Bimodal NTR Option Using TransHab

Another option to the Reference Mission 3.0 under consideration is the use of a bimodal NTR stage to propulsively capture all payload elements into Mars orbit. This "all propulsive" NTR option provides the most efficient use of the bimodal engines which can supply abundant power to the spacecraft and payloads in Mars orbit for long periods. Propulsive capture into the reference "250 km by 1 sol" elliptical Mars parking orbit also makes possible the use of a standardized, reduced mass "aerodescent" shell because of the lower payload entry velocity (~4.5 km/s) encountered. From this orbit, the triconic aerobrake mass varies by only ~400 kg for a 20 ton increase in payload mass (see Section 3.3.3).

The attractiveness of the "all propulsive" bimodal NTR option is further increased by the utilization of the lightweight, inflatable "TransHab" module discussed in Section 3.1. The substitution of TransHab for the heavier, hard-shell habitat module introduces the potential for propulsive recovery of the Earth Return Vehicle in Earth orbit and its reuse on subsequent missions. TransHab use also allows the crew to travel to and from Mars on the same bimodal transfer. In Mars orbit, the crew transfer vehicle rendezvous with the "unpiloted" habitat lander which is now delivered as a cargo element by the bimodal stage. The absence of crew from the bimodal habitat lander eliminates the need for outbound consumables and engine crew radiation shields and allows it to carry off-loaded surface habitation and science equipment previously carried on the cargo lander.

A three-dimensional image of the bimodal transfer vehicle used on the piloted mission is shown in Figure A5-8. The TransHab is ~9.7 meters long and inflates to a diameter of ~9.5 meters. Its total mass is ~24.3 metric tons which includes the crew and their consumables. The total length and initial mass of the piloted transfer vehicle is ~54 meters and ~141 metric tons, respectively. A smaller, "in-line" propellant tank is used on the bimodal transfer vehicles that deliver the ~46 metric ton habitat and ~54 ton cargo landers into Mars orbit. The habitat and cargo transfer vehicles are ~56 meters long and have a LEO mass of ~129 metric tons and 144 metric tons, respectively.

5.3.3 "LOX-Augmented" NTR Option

An enhanced NTR option, known as the "LOX-augmented" NTR (LANTR), is presently under study by NASA which combines conventional LH_2-cooled NTR and supersonic combustion ramjet (scramjet) technologies. The LANTR concept utilizes the large divergent section of the NTR nozzle as an "afterburner" into which LOX is injected and supersonically combusted with reactor preheated hydrogen emerging from LANTR's choked sonic throat-- essentially "scramjet propulsion in reverse." By varying the oxygen-to-hydrogen mixture ratio (MR), the LANTR engine could potentially operate over a wide range of thrust and specific impulse values while the reactor core power level remains relatively constant. For those missions where volume (not mass) constraints limit bimodal stage performance, LANTR propulsion can help to increase "bulk" propellant density and total thrust output, while decreasing the engine burn times. LOX augmentation would be particularly beneficial during the TMI burn to reduce gravity losses. Following this maneuver, the spent "in-line" LH$_2$ tank and a small LOX tank attached to it could be jettisoned as a single unit. On all subsequent burns, the LANTR engines would operate on only LH$_2$ (MR = 0). Cold flow experimental injector tests and reactive computational fluid dynamics analyses are currently underway at NASA Lewis Research Center in preparation for future hot flow tests aimed at demonstrating concept feasibility.

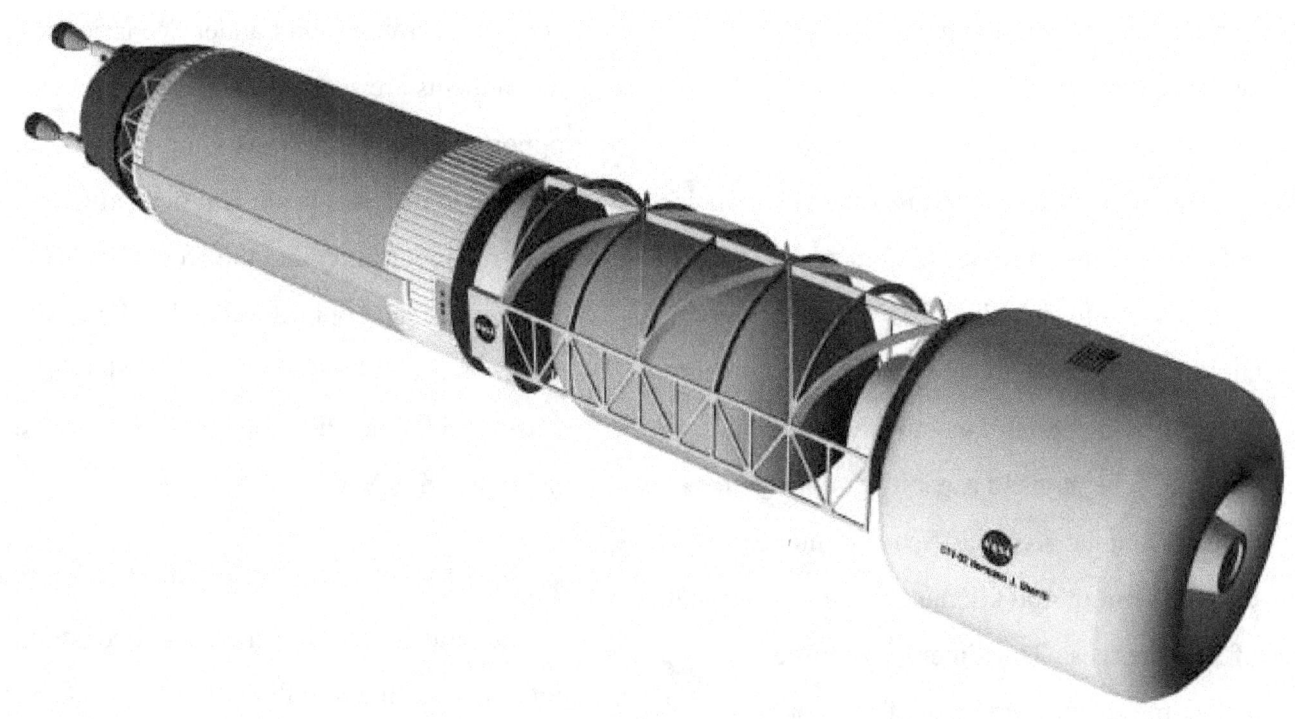

Figure A5-8 All Propulsive Bimodal NTR Carrying TransHab

5.3.4 Three-Magnum Scenario

Point of Contact: Andrew Petro/JSC

During the Spring of 1998 a special design study was conducted to define the elements, mission content, and technology required to accomplish a human Mars mission which could be accommodated for launch within the mass and volume capacity of three heavy-lift launch vehicles. The reference launch vehicle used in the study was the Magnum launch vehicle and so this mission concept is referred to as the "Three-Magnum Mars Mission". The design team was directed to employ a solar electric propulsion (SEP) stage for delivering the Mars mission elements to a high apogee Earth departure orbit and to not employ nuclear propulsion for any maneuvers.

This study was unusual in the approach of designing to a fixed constraint for Earth launch mass. The most significant result of the study was the identification of the technology challenges which must be met to achieve the launch mass goal.

The capacity of the Magnum launch vehicle defined for this study was 89.5 metric tons for launch packages which employ the launch shroud as an aeroshell, and 85.5 metric tons for payloads which do not include the shroud as payload. The payload capability quoted is for launch from the Kennedy Space Center to a circular orbit of 400 kilometers at an inclination of 28.5 degrees. The dimensions of the Magnum shroud were defined as an outer diameter of 8.4 meters and a length of 28 meters.

5.3.4.1 Mission Content

The mission defined in this study included a crew of four people, a scientific payload of 1770 kg and two unpressurized rovers with a mass of 650 kg each. The missions were conjunction-class with outbound and inbound transit durations of 180 to 200 days and Mars surface stay times of 520 to 580 days. The elements were designed to accomplish missions in six out of the eight opportunities in the synodic cycle. The other two opportunities would require an additional propulsive stage of approximately 16 metric tons.

Several different mission scenarios were considered and two were documented for the study: a Combination Lander Scenario in which all elements are sent to Mars in a single opportunity, and a Split Mission Scenario in which some elements are deployed at Mars in the first opportunity and the crew travels to Mars in the next opportunity. The Split Mission Scenario is similar to the Design Reference Mission 3.0 whereby propellant for Mars ascent is produced at Mars.

5.3.4.2 Strategies and Technology Challenges

Several strategies were used to constrain the total mission mass with respect to the Design Reference Mission and to achieve the launch mass target.

- Crew reduced from 6 to 4 persons
- Initial departure orbit apogee raised from 39,000 km to 120,000 km
- Hydrogen fuel is used for all maneuvers.

In addition, several technology development challenges were identified as necessary to achieve the launch mass target.

- Structures, tanks, and aeroshells with a reduction in mass of up to 50% over current technology
- High performance power generation system for space and surface operations (100 kg/kWe)
- Long-term hydrogen storage with near zero boil-off for up to four years
- Lightweight chemical propulsion engines with a specific impulse of 480 sec.
- Deployable solar electric propulsion system with a megawatt-capacity solar array

5.3.4.3 Combination Lander Scenario

This scenario is illustrated in Figure A5-9 and the launch packages with element masses are shown in Figure A5-10. Figure A5-11 is a three-dimensional drawing of the Combination lander concept as it would be deployed on the surface. This lander includes the crew module for descent and ascent along with the surface habitat.

5.3.4.4 Split Mission Scenario

The Split Mission Scenario is similar to the Design Reference Mission scenario but it includes all of the strategies and technology challenges mentioned above. The major differences in this scenario are 1) the pre-deployment of the return vehicle in Mars orbit, 2) pre-deployment of the ascent vehicle on the surface of Mars, 3) the production of propellant on Mars, and 4) the use of methane rather than hydrogen for Mars ascent. The scenario is illustrated in Figure A5-12 and the launch packages and element masses are shown in Figure A5-13.

5.3.4.5 Summary

By incorporating the aggressive technology goals, two mission scenarios were defined which could meet the three-Magnum launch mass and volume constraint. It should be noted that each scenario also requires a Space Shuttle launch at the beginning of the mission to deliver the crew and their high-Earth orbit taxi and also a Shuttle mission at the end to recover the crew in low Earth orbit. This three-launch strategy is reliant on the key technologies described previously. An effort is currently underway to better understand the difficulty of the technology challenges as they compare to current state-of-the-art, the risks associated with these technologies, development costs, and the architectural impacts of potential technology fall-backs if it is believed that the technology development cannot be completed as needed.

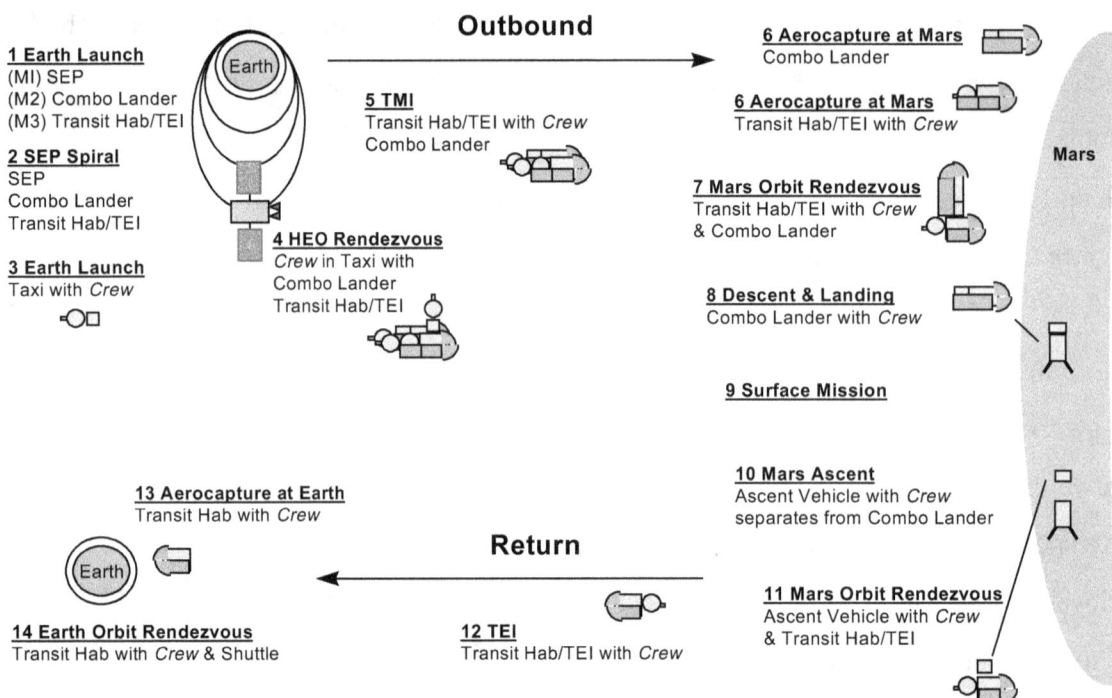

Figure A5-9   Three-Magnum Combination Lander Scenario

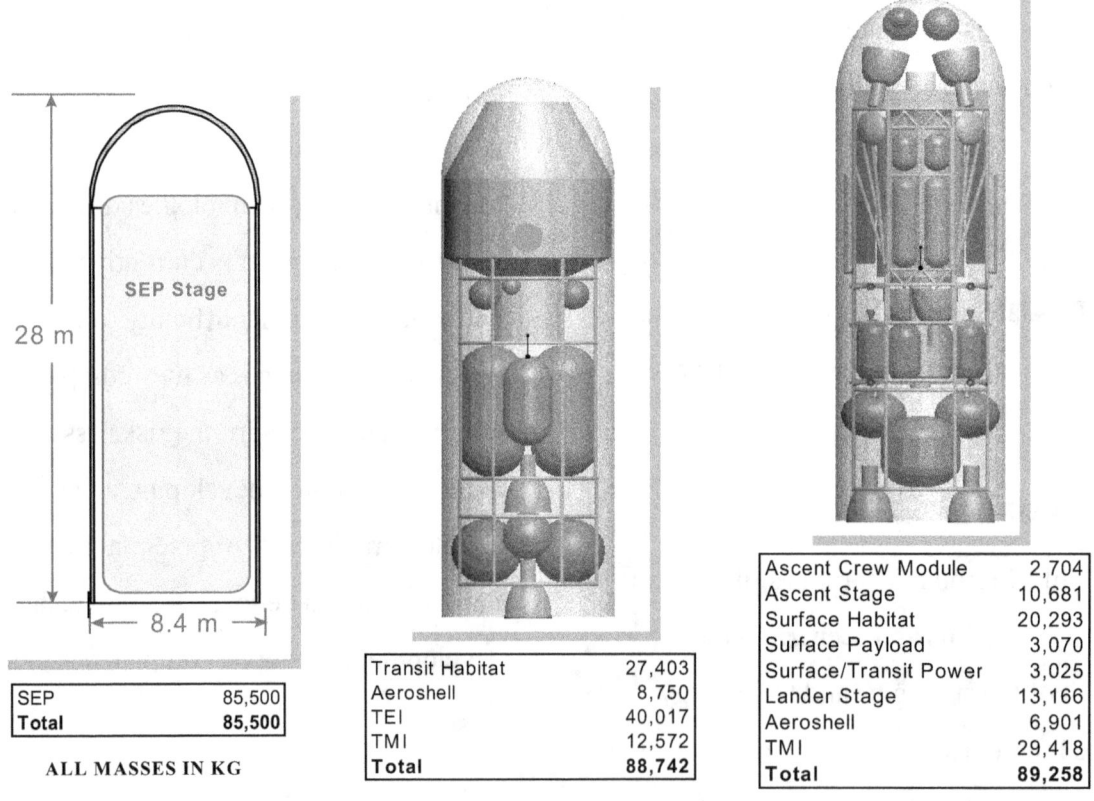

Figure A5-10   Three-Magnum Combination Lander Launch Packages

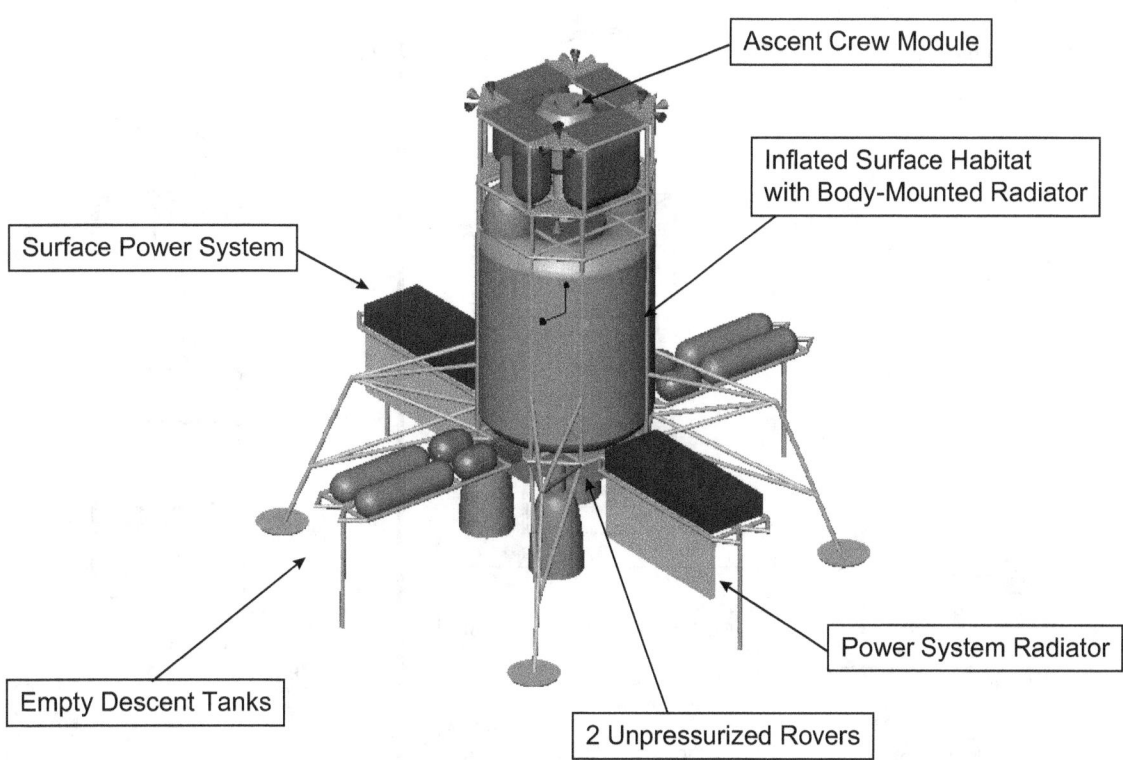

Figure A5-11 Combination Lander Concept on Mars Surface

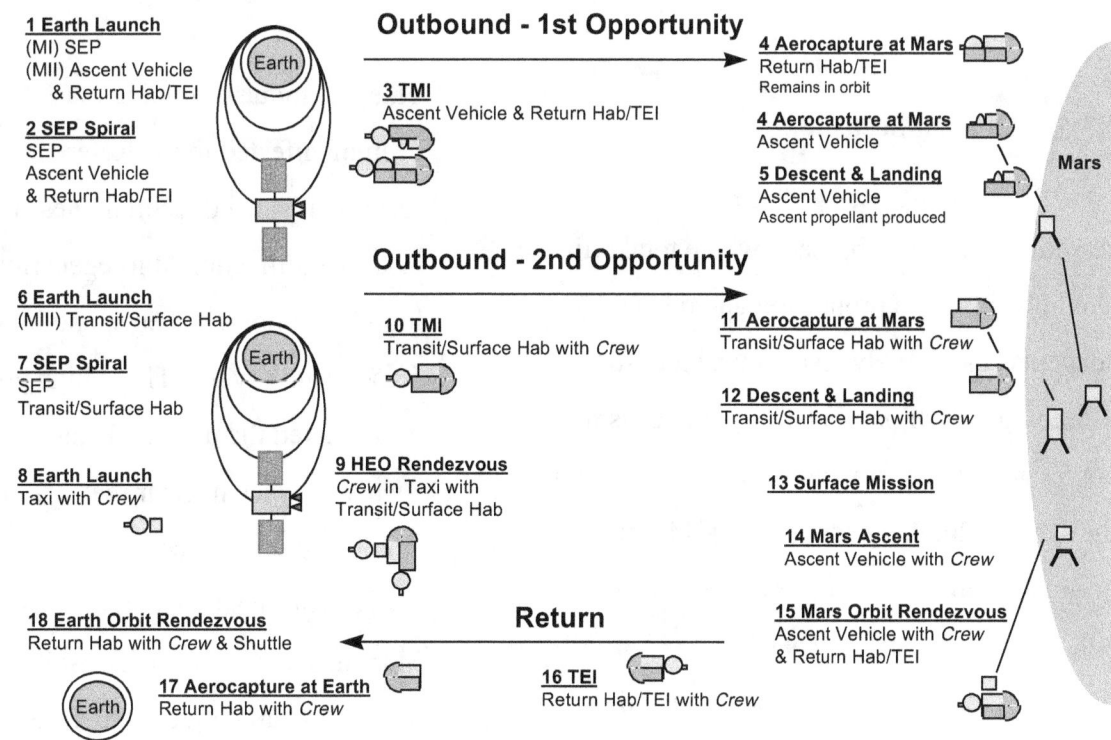

Figure A5-12 Three-Magnum Split Mission Scenario

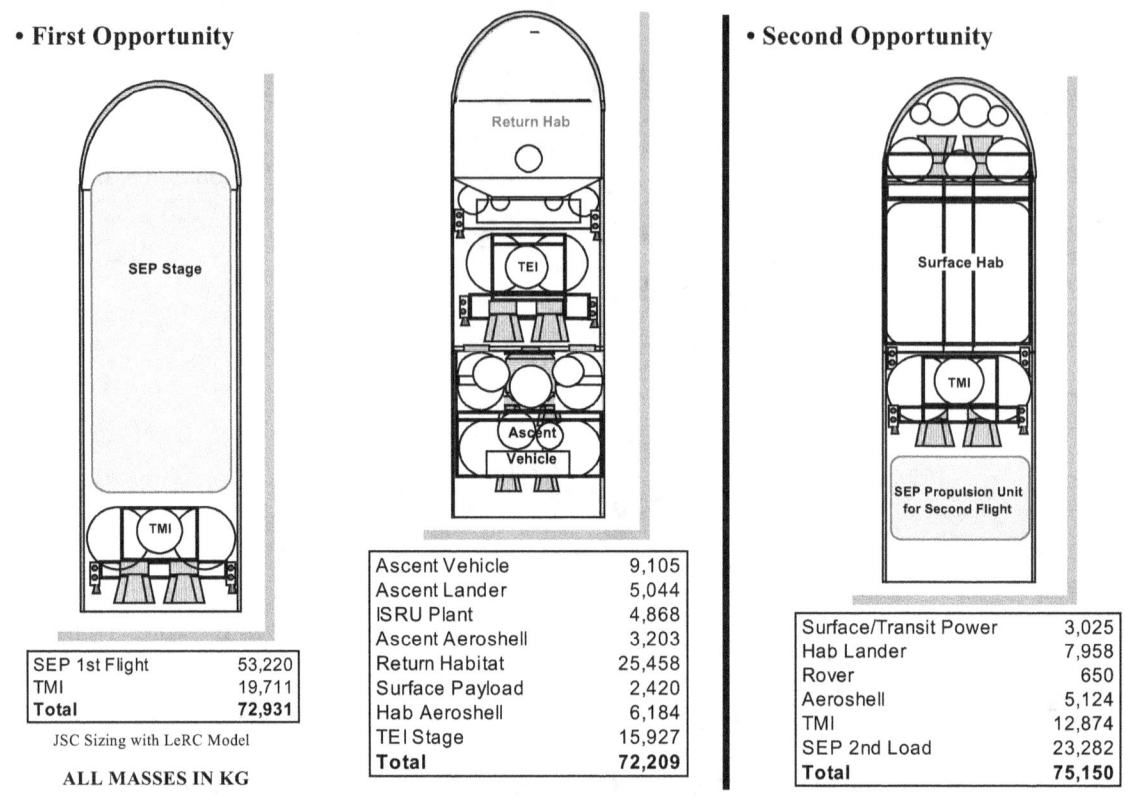

Figure A5-13 Three-Magnum Split Mission Launch Packages

5.3.5 All Solar Scenario

Another alternative strategy under consideration is an approach where total reliance would be place on propulsion and power concepts based solely on chemical and solar technologies. Of particular importance is the power generation strategy which has relied on the same technology base (SP-100) from the original reference through Reference Mission 3.0. This power strategy has been one of providing a robust power generation and storage capability to enable significant mass reductions.

This technique of trading mass for power has been manifested in the Reference Mission in the form of advanced technologies such as in-situ resource utilization, bioregenerative closed-loop life support systems, and long-range pressurized rovers. These high power demands necessitated the use of advanced power concepts such as surface nuclear reactors and dynamic isotope power sources.

A major challenge of an all-solar human Mars mission is the lack of solar irradiation at Mars. As can be seen in Figure 5-14 the solar flux at the surface can be as low as 6.5% of that in low-Earth orbit. The reduction of solar flux is due to the distance of Mars from the sun, the presence of the atmosphere, and potential dust storms.

Analysis of this all-solar mission approach is currently under way. Results of this study will be included in the next update of the Reference Mission.

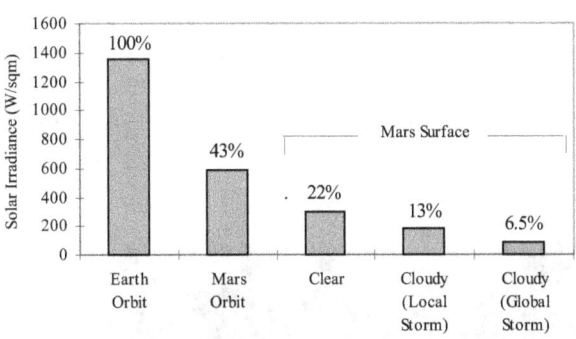

Figure 5-14 Solar Irradiation At Mars.

Analysis of an all-solar approach will include:
- Developing a mission approach where the surface element power needs can be reduced to the lowest level possible
- Understanding the sensitivities of advanced solar cell technologies
- Analysis of solar power generation system setup and maintenance, such as cleaning due to dust accumulation
- Analysis of the impacts of elimination of advanced technologies (in-situ resource utilization, long-range rovers, food production, etc.) on the overall mission approach, including risk.

A6.0 Continuing Work

A "reference mission" is a continual work in progress, provided to the space community to critique and build upon. Future human exploration analysis activities will focus on exploration targets such as the Moon, Asteroids, or other destinations beyond Earth orbit. Further addenda will be published which document these changes to the mission, changes which will undoubtedly be made before it becomes a reality. Through this process, the Exploration Team can nurture the design of human exploration missions which offer a safe and economical strategy for taking the next step in humanity's exploration and development of space.

www.ingramcontent.com/pod-product-compliance
Lightning Source LLC
Chambersburg PA
CBHW081739170526
45167CB00009B/3875